MEDIEVAL MODAL LOGIC & SCIENCE

Augustine on Necessary Truth & Thomas on Its Impossibility without a First Cause

Forewords by

Dr. David Lamb, School of Medicine,
University of Birmingham, Great Britain
and
J. Roland Ramirez, Duquesne University (Ph.D.,
Institut Catholique de Paris)

Robert Trundle

University Press of America,® Inc.
Lanham • New York • Oxford

University Press of America,® Inc.
4720 Boston Way
Lanham, Maryland 20706

12 Hid's Copse Rd.
Cumnor Hill, Oxford OX2 9JJ

Printed in the United States of America
British Library Cataloging in Publication Information Available

Library of Congress Cataloging-in-Publication Data

Trundle, Robert C.
Medieval modal logic & science : Augustine on necessary truth & Thomas on its impossibility without a first cause / Robert Trundle ; forewords by David Lamb and J. Roland Ramirez.
p. cm.
Includes bibliographical references and index.
1. Modality (Logic)—History. 2. Augustine, Saint, Bishop of Hippo—Contributions in modalilty (Logic) 3. Thomas, Aquinas, Saint, 1225?-1274—Contributions in modality (Logic) 4. Philosophy, Medieval. I. Title. II. Title: Medieval modal logic and science.
BC199.M6T69 1999 160'.9—dc21 99—21904 CIP

ISBN 0-7618-1398-5 (pbk: alk. ppr.)

™ The paper used in this publication meets the minimum requirements of American National Standard for Information Sciences—Permanence of Paper for Printed Library Materials, ANSI Z39.48—1984

For
Janeanne Wilcox Trundle

Contents

St. Thomas: Modalities and Causality

Foreword Comments by Dr. Lamb

Drawing from his vast knowledge of the history of philosophy and philosophy of science, Robert Trundle demonstrates how St. Augustine and St. Thomas engaged with the problems of their day and how their reasoning is still relevant. Consequently, this book is not merely about these major historical figures. It addresses contemporary issues such as cultural relativism and the legacy of a Humean-Kantian-positivist tradition. The book engages with the questions "Why is there something rather than nothing?" and "Whence comes truth if not from metaphysics?" In dealing with them, Trundle shows it is not merely euphemistic to say that many modern ideas are not as original as it is often held. The ideas beg for a medieval reasoning, related to metaphysics, for their coherence.

Accordingly, the first section, on Augustine, is not only a historic overview but challenges many theoretical assumptions of modern philosophy which led to complex problems. For example, by relating Augustine to a simple Aristotelian modal logic, this section articulates a reasoning of scientific common sense. Augustine's philosophical theology, in this manner, calls into question many misguided divisions between science and extra-scientific influences, from ordinary experience to theology, on scientific thinking. Trundle demolishes the myth that Augustine represents a form of anti-scientific Christian Platonism. He went far beyond this Platonism in countering verificationist modes of thinking which, argues Trundle, have spread from a positivist philosophy of science to morality.

How can Augustinian theology fare in a world dominated by cultural relativism and a verificationist reasonableness? The fact that verificationism is self-refuting, notes Trundle, seems to have had little impact on those who employ it against religious and moral truth claims. Surely, to some degree, he suggests, this is indicative of intellectual dishonesty. In the course of appraising Augustine's philosophy, with attention to Wittgenstein towards the end, Trundle extracts a theory of scientific truth and a notion of freedom which can be contrasted with verificationism and relativism.

These themes are systematically pursued in the second section on Thomas. It begins with his five arguments for the existence of God which anti-metaphysical philosophers of the twentieth century have dismissed as academic pedantries. But Trundle's comparison of Thomas with Wittgenstein reveals fundamental similarities. For example, Wittgenstein's *On Certainty* eroded the analytic-synthetic distinction, which was the cornerstone of positivist philosophy, and has parallels with Thomas' erosion of a radical faith-reason distinction. What is fundamental to both philosophers is the notion of "unclear boundaries". In this context, Trundle refers to W. H. Walsh's examples of categorial mistakes.

One version of such a mistake involves a situation where one is a passenger in a car. The car stops and, when asked "why?", the driver replies that there is no reason, it just stopped. This answer expresses a peculiar kind of nonsense; it is not an empirical error, like wrongly claiming to have run out of fuel, nor is it a logical error. It is categorial nonsense in the sense that "stopping for no reason" offends against epistemic parameters for conducting empirical enquiry. This kind of example indicates how modal truth is located somewhere between logical and empirical truth. Trundle locates these strong modal truths in Thomistic thought and applies them to a range of case studies from geology and medical science to biology and physics. This is *first rate* philosophy of science. And it bears stunningly on, what may be called, a "scientific proof of God" which has major implications for society.

David Lamb, Ph.D. Department of Biomedical Science
and Biomedical Ethics, School of Medicine, University
of Birmingham, Edgebaston, Great Britain

Foreword Comments by Dr. Ramirez

I enjoyed reading this book especially because it shows that, latent in Augustine and Aquinas, there are all sorts of medieval ideas which anticipate what contemporary scientists as well as philosophers have found of value in their own proper domains. These domains are almost invariably presumed to be "emancipated" from medieval concerns for obsolete objects. Such objects, incongruous with rational inquiry, supposedly include metaphysics, God, free will, religion, and immortality.

Whatever might be said about Robert Trundle's work on the reasoning of Augustine and Thomas, it is refreshing to read a book where the author is at home in so many different disciplines. To be conversant in but one of the many areas that this book encompasses — Augustine, Aquinas, modal logic, philosophy of science, epistemology, metaphysics, natural theology, logical positivism, linguistic analysis, plus Greek, medieval, modern, and contemporary philosophy generally, including authors such as Wittgenstein and Heidegger —is already no mean accomplishment.

One notable strength of this work, in this context, is that it shows the current relevance of Augustine and Aquinas to the whole of philosophy and science, and to contemporary science in particular. Trundle brings to bear on this science what he calls Augustine's "existential essentialism" and its contribution to a coherence of "truth" in terms of the scientist's free choice of will. And he does these things by his own elaborate amplifications of modal logic.

Another strength, perhaps Trundle's characteristic contribution, is that he establishes the relevance of medieval thinkers to the tools, language, and presuppositions of the philosophy of science as well as to the logic, methodology, and pertinent givens of science itself. That relevance includes the broader areas of metaphysics, religion, and non-logical realms of human enrichment which the narrow post-Enlightenment focus on "reason" and empirical verification tended to abandon. The abandonment, particularly, impoverished an *entirety* of human truth which bore on science and scientific methodology themselves.

According to Trundle, "Thomas strengthened a realist construal of scientific truth by metaphysico-modalities. Anti-metaphysical modern philosophies undercut scientific truth, often held to be ideal knowledge, as well as any truth in morality and politics." These developments, in no small measure, are instrumental in creating the peculiar *angst* of modern man and in bringing about the restrictive political ideologies which are unique to modern society.

But even if Trundle said nothing about either Augustine or Aquinas, whose ignored insights set the stage for a societal nihilism, his book constitutes a knowledgeable and virtually internal critique of inherited assumptions and hidden fallacies. For the latter lie dormant in the axioms, epistemologies, and methodologies accepted by modern scientists as well as by philosophers of science, and his critique employs their own terminologies and methods. The many questions raised by Trundle are matters from which *any* present-day thinker or educated person concerned with the present, or the future, of any philosophy or of any science could seriously benefit.

J. Roland Ramirez (Ph.D., Institut Catholique de Paris)
Professor of Philosophy, Duquesne University
Pittsburgh, Pennsylvania, USA

Author's Preface

This book probes complementary modal insights of Saints Augustine and Thomas that bear on our most knotty epistemological problems. Herein, it is assumed that a contemporary rediscovery and development of modal logic, often with an esoteric focus, should not exclude fresh inquiry into how its medieval forerunner may contribute to solving the problems. It seems prudent to entertain the insights, when they accord with ordinary experience and language, even if they at times seem at odds with theoretical developments of modal systems that are themselves subject to disputed interpretations.

In largely leaving the interpretations to others, with an exception of noting modal approaches in response to possible objections, the insights begin with ignored Augustinian themes: Matter is infused with thought, a denial we are conscious of thought is senseless, and we are incontrovertibly conscious of a free choice of will necessary for moral as well as scientific truth. And contrary to a supposed religio-antiscience Platonism, Augustine has intriguing ideas on science in terms of a *de dicto* modal logic. This logic goes beyond Aristotle, reflecting a *disregarded Aristotelianism*, and even suggests an impossible falsity of well established scientific laws.

The laws bear on Thomas' reasoning to God, understood as a First Cause, wherein it is impossible there is a world of second causes when there is no First Cause. This reasoning gives expression to a scientific proof of God in the sense, precisely, that certain exceptionally well-established scientific laws strictly imply that the laws cannot obtain when there is no

God. To put it another way, the impossibility may be recast as a necessity and the laws understood as modalities: To accept modal scientific laws which express relations of second causes, such as a thermodynamic law, is to accept that necessarily if there is no First Cause, these laws do not obtain. Since the laws do obtain, there is a First Cause called "God".

Moreover, this proof of God involves an idea of causality that, if modally true, permits an inference to the truth of a modern causal principle. This inference is momentous if the principle's current metaphysical status of not having known truth excludes any "truth" afforded by scientific inquiry. The inference also bears on modern anti-metaphysical philosophies if they arose in response to the question "Whence comes truth if not from metaphysics?".

Do Thomas' metaphysics presage modal insights in Wittgenstein's *On Certainty* and their neglect in Heidegger's *Metaphysics*? The question of why there is a world rather than Nothing evoked the sarcasm that scientists want to know "nothing of Nothing". Perhaps, scientific experience, even if implicit because of its patency, indicates that there never was Nothing by virtue of a modal impossibility of there being a world when there is no uncaused First Cause.

Besides being demanded by a cosmic causal dependence based on experienced phenomena in science, does this *voluntary* Cause explain a free will that makes coherent moral praise and blame? As a Creator of Nature, does it permit both reasoning from natural theology to science and common-sense inferences from scientific descriptions of our nature to its fulfillment without a Naturalistic Fallacy?

Given that ideas have consequences, the question is not academic. It bears on defensible values, if not a survival, of modern societies. Though idealizing a rationality of science, philosophers in these societies have been unable to either tenably speak of scientific "truth" or relate it to ethics. Any tenable talk of the "truth" is in virtue of a rediscovered modal reasoning. But the latter has ignored a presupposed causal reasoning related to a First Cause. In failing to relate the Cause to scientific truth, there is still no viable ethics in modern thought that bridges science and politics. Consequently, a politicization of truth has ensued, unique to modern society, that underscores an urgency of the task at hand.

Acknowledgments

In accord with an old poem "The road to success is to err and err but less and less," I am indebted to many eminent scholars for a cultivation of my ideas. The scholars abroad most significantly include Fr. Daniel Kroger, PhD. As guest editor of *Sophia: An International Journal of Philosophy* under Managing Editor Dr. Rolando Gripaldo, he added to my ideas in "A Scientific Proof of God" (vol. 28, no. 2, 1999). It was his "suspicion that the Aristotelian elements found in Augustine have their roots in both Middle Platonism and in Cicero's 'lost work' the Protrepticus..." Other scholars include Professor van Bendegem, Editor of *Logique et Analyse* at the Belgium Center for Research in Logic; Professor Kris Deschower, Editor of *Res Publica* of the Leuven Politologisch Instituut; Fr. Jose Oroz Reta, Editor of *Avgstinvs* at the Universidad Pontificia De Salamanca in Madrid, Spain; and Professor M. Sorondo, Editor of *Aquinas* at the Pontificia Univerita Lateranense.

I owe special gratitude to Professors Lionel Ponton and Thomas De Koninck, Editors of *Laval Theologique et Philosophique* at Laval University in Quebec, who provided a prestigious forum for my ties of traditional theology to modern science when my efforts seemed at odds with the *idee recue* of many American journals. And my ideas on science and Greek philosophy were allowed to be honed, in a similar way, by the support of David Lamb at the University of Birmingham in England, Editor of the Avebury Series in Philosophy for Gower Publishers. He kindly oversaw several of my projects: *Ancient Greek Philosophy* (1994) and an article for *New Horizons in the Philosophy of Science* (1992).

There were scholars in the United States whose criticism of my submitted work was so insightful that it made me more grateful than the work's acceptance or whose article invitations, in esoteric areas, stimulated new interests that also fortified old ones. The scholars most notably include Professor Peter Longo, Editor of *Idealistic Studies;* Professor G. Richard Dimler, S.J., Editor of *Thought;* and Professors Michael Heller, William Stoeger, S.J., and Josef Zycinski who are both Editors of *Philosophy in Science,* affiliated with the The Pachart Foundation (Tucson, Arizona).

Also, I am beholden to several distinguished philosophers who have areas of scholarship that range from medieval philosophy to the philosophy of science and modal logic. They had either interesting insights or possible objections to specific points. My responses are noted in the body of this work and referenced in footnotes. These philosophers, also listed in the bibliography, include Norris Clarke, S.J., Professor Emeritus of Philosophy at Fordham University; R. J. Connell, Professor Emeritus at the University of St. Thomas; Del Ratzsch, Professor of Philosophy at Calvin College, and Jean Paul van Bendegem, Professor and Chair of the Philosophy Department at Vrije Universiteit Brussel.

Finally, I am grateful to the following publishers for the use of quotations: Transaction Publishers, Rutgers -- The State University, New Jersey, for W. Donohue's *The New Freedom;* Wadsworth Publishing Company, Belmont, California, for John Nolt's *Logics;* The Macmillan Publishing Company, New York, for St. Augustine's *On Christian Doctrine*, Tr. with an Intro. by D. W. Robertson, Jr., and for St. Augustine's *On Free Choice of the Will*, Tr. by A. Benjamin and L. Hackstaff with an Intro. by Hackstaff; Penguin Books Ltd., Harmondsworth, Middlesex, England, for St. Augustine's *Confessions,* Tr. with an Introduction by R. S. Pine-Coffin; Image Books, a Division of Doubleday and Company, New York, for St. Augustine's *City of God,* Tr. by G. Walsh, S.J., D. Zema, S.J., G. Monahan, O.S.U., and D. Honan, with a condensed original Forward by E. Gilson and Ed. with an Intro. by V. Bourke; and the Ignatius Press, San Francisco, for passages in St. Thomas Aquinas' *Summa Theologica,* Tr. in 1920 by The Fathers of the English Dominican Province, in *Summa of the Summa,* Ed. and Annotated by Peter Kreeft.

Chapter 1

Augustine: Plato and Aristotelianism

The thesis that an Aristotelianism found its way into Augustine's thought may not be the received view of many philosophical anthropologists. But it would be an absurd methodology to dismiss the patency of Aristotelian concepts before our eyes, in the writings, on the grounds that to the best of our knowledge Augustine did not have access to Aristotle's central works. Here, a revitalized anthropology is not submitted but rather the obvious influence.[1]

Surely, since Augustine died over eight hundred years before Thomas was born, it is no small thing to say he did not merely hold an Aristotelian epistemology before Thomas. He complemented it by the notion that God's omniscient reason is eclipsed, or hidden from, our reason by His omnipotent will. Besides preceding Thomas' epistemology and illuminating ignored elements of Scripture, his insights bear profoundly on modern philosophical developments. Explicable only by the Aristotelianism, the insights include a modal logic whose import for the developments ranges from a Humean-Kantian critical thought to themes of phenomenology. Contributions to the latter, if not to Wittgenstein's analyses, have been acknowledged. However, the nexuses have not been considered in the manner explored herein.

---

1 Besides the scholarship of Anton-Hermann Chroust noted below, see R. Trundle's "St. Augustine's Epistemology: An Ignored Aristotelian Theme & Its Intriguing Anticipations," *Laval Theologique et Philosophique* 50 (1994) 187-205.

Augustine's stunning contributions may have been ignored because he was associated with Plato's a priori essentialism rather than with Aristotle's organismic empiricism. But both Aristotle and Plato presented paradigm philosophical options and an open-minded reading of Augustine's work reflects his mediation between them. This is not to say that a traditional understanding of his Platonism is entirely skewed. Perhaps, it is to say that the understanding may have been simplified for purposes of teaching and that, given a secular neglect of medieval studies until recently, this fact might have been especially ignored in extra-scholastic research.

Since research bears on the integrity of teaching and both are important, an appreciation of Augustine begins with a discussion of the different roles of Plato and Aristotle. Whereas the next chapter discusses ignored Aristotelian modalities related to the laws of thought and to science, later chapters indicate how he contributes to resolving modern problems of "truth" that also bear on society. Consider, now, what may be called the Platonic and Aristotelian options.

Augustine was faced with equally attractive and unattractive options in the philosophies of Plato and Aristotle. Tradition holds that Augustine was influenced by the Roman statesman Cicero (106-43 B.C.) who wrote the *Hortensius*. Recent scholarship suggests that this work was almost a Latinized version of Aristotle's *Protrepticus*:[2]

> As its title reflects, the *Protrepticus* is both a personal address and a general appeal or exhortation, encouraging future generations to choose wisely when committing themselves to a definite way of life. The experiences of St. Augustine attest the efficacy of this appeal.

Though by tradition Aristotle's *Protrepticus* is "the foundation... for the conversion of the young St. Augustine to the intellectual life and thus indirectly to his conversion to Christianity,"[3] the influence goes far beyond this initial primitive Aristotelianism.

---

2 See Anton-Hermann Chroust's *Aristotle: New Light on His Life and on Some of His Lost Works*, Volume II (IN: University of Notre Dame Press, 1973), p. 121.

3 *Ibid.*, p. 121.

Scripture distinguished between an unchanging spiritual reality and a changing physical one. The distinction, together with a Platonic soul (*psyche*) that knew eternal Ideas, made Plato's metaphysics attractive. At the same time, the attractiveness was diminished by the New Testament revelation that all things, including knowledge, would pass away except Love (*agape*). Moreover, in addition to this Love being connected to humility and secular knowledge to pride, the body's sanctity and resurrection were inconceivable on the Platonic paradigm. The paradigm construed the body as a "dark" receptacle that obscured a cognitive "light" of the soul. This is what St. Paul meant, in large measure, when he said in the Greek city of Corinth that Christian doctrine was foolishness to the Greeks: The "Greeks look for wisdom, but we preach Christ crucified: a stumbling block to Jews and foolishness to Gentiles."[4]

Apparently, St. Paul had been in a less evangelical mood when he previously spoke to an elite group of highly-educated Corinthians who were familiar with philosophy. Paul had been sensitive to their philosophical reluctance to grasp the unprecedented historical and spiritual fact of Christ's resurrection. Now, he bluntly speaks about what he takes to be philosophically enigmatic, a stumbling block to human reason. Just over four centuries later, Augustine wrestled with the insights of Platonic and Aristotelian philosophies. Philosophical reasoning done well would not only be compatible with Scripture but be a short step to revelation.

On the One Hand... On the Other...

On the one hand, it was easier for Augustine to hold Aristotle's philosophy. This philosophy did not merely avoid Plato's unChristian notion of a preexisting soul that had eternal knowledge and hinted at reincarnation (*Phaedo* 71e). The philosophy eluded an anomaly that particular things only partly "share in" universal Forms. When the things in question are horses, for instance, Aristotle held that the form is a biophysical essence of "horse" which is embodied fully by each

---

4 I Corinthians 22, New International Version of *The Holy Bible* (1978).

and every perceived horse. For one thing, this thesis was simpler because there was one world rather than Plato's two. For another, an omniscient God would have a role in illuminating the soul's knowledge of forms. Finally, a unity of form and matter implied a unified soul and body. And while Aristotle interpreted their unity to mean they perished together, the oneness might render viable a bodily resurrection.

On the other hand, the notion of resurrection was tainted by an Aristotelian view that permitted only a single "corruptive" reality. And while this reality led to Thomas' cosmological proof of God, it proceeded ad infinitum into the past. In addition to an infinite past conflicting with the creation of the world out of nothing (*creatio ex nihilo*), the Aristotelian ontology excluded a transcendent God.

Faced with equally attractive and unattractive options, Augustine evidently embraced both. The Platonic ontology of two realities explained *(i)* the transcendence of God, *(ii)* an inferiority of corporeal reality, and *(iii)* a sin-laden human condition. This condition reflected the relative impotence of human knowledge and the prideful self-deceptive tendency to exhaustively appeal to the knowledge for both ameliorating moral squalor and obtaining political utopias. It becomes clear that there is a stress on Plato's theory of reality for spiritual insight about God's realm as well as on Aristotle for explaining both observational and theoretical knowledge in the worldly realm. This realm, nevertheless, reveals God's mercy for those who have "eyes to see".

Aristotelian Truth

In leaning on the notion that a form or essence (*essentia*) is embodied by the individual, secular philosophers and scientists may ignore a spiritual reality which begets humility. By succumbing to the pride of worldly knowledge, there is a grave risk of pursuing godless political utopias that, in the name of "progress", incur untold suffering and death.[5]

5 Aristotle did not himself advocate political utopias. "Let us remember...," he cautions in his *Politics* (1264a), that "in the multitude of years these... [utopian ideas], if they were good, would certainly not have been unknown..."

Despite this risk, an Aristotelian epistemology might explain a merciful immanence of God. In being reminiscent of Aristotle's divine *Nous* (Mind), God's goodness and intelligence were manifest in the world. The world is an ordered and predictable realm that, while orienting us to the empirical, cannot be articulated entirely by the bodily senses. In *On Free Choice of the Will* he says we must admit that the "truth of numbers... [has] nothing to do with the bodily senses."[6]

Augustine does not mean that mathematical ideas refer to entities in a "really real world" of Platonic Ideas. He means, among other things, that a mere materialistic theory of perception is limited to perceived *particular* things and cannot explain *universal* ideas. The theory may explain how we observe three circular things but not universal ideas such as "three" or "circle" or "π ".

An account of such ideas, in his *City of God,* involves our having inner lights of intelligence.[7] We have intelligent immaterial souls whose exercise depends on a spiritual God. Thus, we are able to give numerical expression to parts of bodies "we admit, must be *in* a body, however small it is."[8] Taken with other insights noted later, he suggests that bodies have Aristotelian essences of "inner" and "outer" forms whose illumination affords universal ideas. Others include "small". And given that forms are not in a Platonic realm, this term would lose its epistemic import for what really, but relatively, is true about bodies if universals did not really inhere in them.

Augustine is not at odds with some versions of modern scientific realism. Keeping in mind that the Greek word "*techne*" bears on medieval "science" as a continuum from art to physics to geometry, the realism suggests something akin to essential natures: In ignoring empirical predicates "accidental" to a given purpose such as mechanics, science assimilates the essential ones into exact theoretico-universal ideas of relations,

---

6 St. Augustine, *On Free Choice of the will,* Tr. by A.S. Benjamin and L.H. Hackstaff (NY: Macmillan Publishing Co., 1985), p. 36.

7 St. Augustine, *City of God,* Tr. by G. Walsh et al (NY: Doubleday & Co. Inc., 1958), pp. 256, 526.

8 St. Augustine, *On Free Choice of the Will,* p. 55, emphasis added.

processes, and properties.[9] A property does not change in given domains, *e.g.* "mass", but rather superseding theories which increasingly reflect how it really *is*. A theoretical area "$A_o = 6a^2$", by analogy, is an exact geometrico-universal notion inferred from observed cubes, where "accidental" predicates such as types of substance are disregarded.

When instantiated with a given value, "A_o" may lead to a veritable idea of "small area" even if it does not inhere in bodies. A cubic body is small relative to others without involving Plato's realm of universals. Where does Augustine ever speak of universals in such a realm? In short, we have an inner intelligence, which depends on divine illumination, that reflects an Aristotelianism consistent with his Doctrine of Illumination and modern thought: A dependence on the sun for observing bodies from which there are theoretico-universal inferences dependent on God:

GOD))) ILLUMINATES *ONE* UNIVERSAL CUBIC IDEA

$$"A_o = 6\,a^2"$$

SUN))) ILLUMINATES *MANY* OBSERVED CUBES

Doctrine of Illumination

Platonic Reality

Though Augustine posits an inner light of our intelligence that relies on divine illumination, he seeks to reveal a tendency to confuse our intelligence with a god-like wisdom. Despite a limited wisdom to use our intelligence for good, the "good" may be obscured by pride. Even Aristotle's "proper pride" falls short of the humility needed for truth about a spiritual reality. A distinction of this reality from the corporeal world reflects the Platonic influence. In *On Free Choice of the Will*, Augustine notes Solomon's search for God's wisdom: "I and my heart have gone round to know ... wisdom and number."[10] Other distinctions,

---

9 Cf. Stephen Korner, *Experience and Theory: An Essay in the Philosophy of Science* (London: Routledge & Kegan Paul, 1966), pp. 163-178.

10 *Ibid.*, p. 57 and Ecclesiastes 7: 26. "Wisdom" may be a word for God.

such as knowing rules of logic and and true propositions, are made in *On Christian Doctrine.*

Thus Augustine contrasts his doctrine of wisdom, that emphasizes humility, to those who pridefully boast when they know trivially true "rules of valid inference as if they had learned the truth of propositions."[11] It is better to know true propositions than to reason vacuously or sophistically by rules of inference. Whereas the rules are tautologous since their denials are self-contradictory, nontrivial truth reflects the way in which *is* reality.

For example, whereas reality may be the truth-condition for true substitution instances of $p \longrightarrow q$, the inference rule *Modus Ponens* "$p \longrightarrow q$ and p, therefore q" has a self-contradictory denial since all substitution instances are false and nondenial ones true. Hence, knowing the true proposition "there is a resurrection of the dead," notes Augustine, is better than knowing that "it follows from... there is no resurrection that 'then Christ is not risen'."[12]

In contrasting this *modus-ponens* inference to the truth of propositions, he seeks to *(i)* show that reason needs to be based on our experience of human nature and Nature, and *(ii)* reveal the strongest sort of non-trivial truth. This truth is about a spiritual reality similar to Plato's invisible realm since it is immutable, as are this world's essences, but eternal as well.

When one can say what a thing *is* without qualification because it is eternal and unchanging, it is more real than a thing unchanging but temporary. When an essence is temporary, even if temporarily unchanging, truth about it is qualified by time. And time is itself a created creature. The spatio-temporal realm embodied by essences or forms will come to an end, as the Scripture says, even if it is infused with something akin to a divine intelligent Form.

Called by Aristotle a "Nous", God is not reduced to such a Form. But to say there are *many* "forms" is to say prima facie, on the heels of a "One over Many" of Plato as well as Aristotle,

---

11 St. Augustine, *On Christian Doctrine*, Tr. by D.W. Robertson, Jr. (NY: Macmillan Publishing Company, 1988), p. 70.

12 *Ibid.*, p. 70.

that there is *one* Form. As nature's intelligence, it renders intelligible the cosmos and interrelates essences that are intrinsic. Augustine refers, as discussed later, to a divine "intrinsic power... fill[ing] heaven and earth" and to an "intelligent nature" from which stem "inner forms".

No pantheism here. Precisely, forms are studied by formalized sciences; all things are not mere modes of God; and He is transcendent as well as an Efficient and Final Cause external to the cosmos. Further, to object that his cosmic query "Who made it? How? and Why?" pertains only to the Trinity, *e.g. De Trin. vi.*, is to assume the very religio-antiscience Platonism in question. Finally, given his stress on divine will as well as reason, *via* a Final Cause, it is also reasonable to hold that created forms are infused into the cosmos for a divine end.

Whereas any Aristotelian end was unknown, Augustine pioneered the first philosophy of history, inspired by theology, from birth to death to when death reigns no more. Akin to old clay condemned originally (*"massa... quae originaliter est tota damnata"*), with Christ's new clay central to history, death is overcome as we are molded into eternal vessels (*"prius... est reprobum... posterius vero probum"*):

CREATION: FALL > HUMAN BIRTH AND DEATH >

DEATH OF CHRIST > CHRIST'S RESURRECTION >

OUR POTENTIAL RESURRECTION:

RESURRECTION, JUDGMENT,

AND END

History's Revealed Panorama

This panorama returns us to an inner light of intelligence that includes a lower wisdom with its source in the revealed wisdom of God. There is not a changing relativistic "conventional wisdom", where what is wise at one time may be unwise at another — so that it may be both true and false that any given idea is wise. Rather, there is only one higher wisdom of God. In illuminating our ideas of history and morality as well as logic and science, God is like one sun by which there are many illuminated objects :

> Just as the objects which men see in the sunlight and choose to enjoy are many..., the light in which the sight of each man... sees and holds what he enjoys is one; so... the very light of wisdom... is one [godly] wisdom common to all wise men.[13]

Reality and Truth: Wisdom

Having the name, among others, of "Wisdom", God is ultimate Reality and Truth. Without obviating the notion that the created world is a truth-condition for truth which imparts our distinction of "truth" from "reality", what distinguishes wise men from foolish ones is that the latter elevate an inner light to an absolute one wherein, in effect, they deem themselves gods. By contrast, a humility of wise men has been perennially reflected in their deference to the wisdom of God. In pondering the fate of America during its Civil War, Abraham Lincoln said in one of his heartfelt letters: "By the help of an all-wise Providence, I... expect the continuance of your prayers for a right resolution of our national difficulties".[14]

In difficult times, we tend to seek God's wisdom. This is why, as Augustine suggests, the Scripture reveals that *sorrow* is better than laughter. In echoing James 4:9 to "Change your laughter to mourning and your joy to gloom," Augustine says "Have pity on me, O Lord, in my misery !... In prosperity as the world knows it there is twofold cause for grief."[15] Even in the ancient Greek *Agamemnon*, the grief noted by Aeschylus is a short step to revelation:[16]

> He who learns must suffer. And even in our sleep pain that cannot forget, falls drop by drop upon the heart, and in our own despair, against our will, comes wisdom to us by the awful grace of God.

---

13 St. Augustine, *On Free Choice of the Will*, p. 59-60. Emphasis added.

14 R. P. Basler, Ed.,*The Collected Works of Abraham Lincoln, Vol. V: 1861-1862*, (NY: Rutgers University Press, 1953), pp. 215.

15 St. Augustine, *Confessions*, Tr. by R. Pine-Coffin (Middlesex, England: Penguin Books Ltd, 1984), p. 232.

16 See A. Schlesinger, *Robert Kennedy and His Times* (NY: Ballantine, 1978), pp. 666-669, where after the assassination of his brother with an ensuing sorrow and pessimism, Robert Kennedy memorized these lines of the *Agamemnon*. He came to see "that the mystery of suffering underlay the knowledge of life."

Augustine's pessimism about recourse to wisdom seems especially relevant to America. Since its divorce epidemic in the 1970s, with a quantum leap in social pathologies, there have been unparalleled aspirations for self-gratification and liberation that center on attacking marriage. This has led to assaults on the Church.

The "church, in particular, is scorned," notes social scientist W. Donohue. "After bashing Saints Paul and Augustine, the Jesuits get it next, followed by a tirade against the Puritans."[17] Traditionally, marriage invites sober concern for the blessings of children, fidelity, and an indissoluble sacramental union (*proles, fides, sacramentum*). The union rests on a sacrificial commitment of unconditional love and not on passively feeling love or on self-fulfillment in terms of changing social agenda.

Spreading like an epidemic, the agenda beckons us to "move with the times" and to liken changing conventions to wisdom. Given that revealed wisdom is experienced palpably,[18] and without being quaint, a foolishness of conventional wisdom begs to be noted by way of classic movies that venture to unveil what is "love". Whereas *Love Story* and *Wizard of Oz* contain the lines "Love means never having to say you're sorry" and "Remember, it's not how much *you* love but how much you are loved,"Augustine stressed the exact opposite.

Anyone who has struggled to sustain an intimate relationship has experienced the paralyzing pride that prevents saying "I'm sorry". Nonetheless, it must be said in response to a virtual impossibility of not being frequently selfish. Selfishness is a manifestation of a human condition that craves an inner peace it cannot have without unconditional love. And with respect to valuing how much one *is loved* over how much *one loves*, Jesus revealed what is love with the words "Father, forgive them, for they know not what they are doing" after being left to die for doing no wrong. There is no love without forgiveness.

---

17 W. Donohue, *The New Freedom: Individual and Collectivism in the Social Lives of Americans* (NJ: Transaction Publishers, Rutgers —The State University, 1990), p. 98.

18 St. Augustine, *Confessions*, p. 231: "You are Truth... present where all seek counsel of you."

In presupposing humility, forgiveness frees us from a pride by which we deceive ourselves and can be manipulated by others. The words of Jesus, "I thirst", Augustine reminds us, were held foolishly to express mere bodily need. By grace, because of faith, the thirst is revealed as a vain desire for love from those who hooted and cheered at his suffering but who could not change the truth he personified.

The foregoing is not a sanctimonious polemic since, as Augustine well knew, Christ did not come for the righteous but rather for the sinner. Moreover, the insights are related poignantly to neglected modes of experience that are not mere physical sensations and that bear on epistemology. There is not only a perennial difficulty of seeing the epistemic speck in our brother's eye and not the log in our own but, paradoxically, prideful demands that reality as a truth-condition conform *a priori* to our varying conceptions. Could compelling criticism of the conceptions, by way of Augustine's philosophical theology, be explained unless the theology reflected an ignored truth about human nature and Nature?

Intriguingly, the question about Nature not only invites renewed attention to theology but to a common-sense modal reasoning.[19] Since the reasoning of modern science is typically regarded as a paradigm knowledge-yielding enterprise, it is simply a matter of intellectual honesty — if not of wisdom — to consider some extraordinary research on the question by pre-eminent members of the scientific community. They do not merely challenge the dogmatic assumption that theology has no warranted role in contributing to the philosophy of science and scientific truth. Their research indicates that the truth has, in point of fact, been advanced by theology throughout the history of science.

---

19 *Cf.* R. Trundle, "Modalidades aristotelicas de san Agustin: En memoria de Fr. Jose Oroz Reta," *Review of Augustine* (Avgvstinvs) 42 (1997) 13-40.

Chapter 2

Aristotelian Modal Thought

Before examining how Aristotelian modalities in Augustine's thought may bear on general philosophy and the philosophy of science, in stunning but ignored ways, an important point needs to be stressed. Though science is typically regarded as ideal knowledge, it is a human endeavor that is not isolated prima facie from Nature being understood aesthetically and theologically.

Thus although it is often assumed that extra-scientific influences on scientific inquiry undermine its integrity, three eminent scholars who reject this assumption are Harvard University physicist G. Holton, lecturer in philosophy of science at the University of London H. Chang, and Max Planck Institute physicist E. Jurkowitz. They note that the influences have advanced science even if this fact "rarely survives in the published record" and that "personal thematic presuppositions of various sorts are found to be essential... in major advances throughout the history of science."[1]

For example, K. A. Muller's codiscovery of high temperature superconductors was guided by a religio-philosophical symbol inspired by Wolfgang Pauli who himself published an essay on influential archetypal conceptions of "Kepler in a book coauthored with psychoanalyst Carl Jung (Pauli 1952)."[2]

---

[1] G. Holton, H. Chang, and E. Jurkowitz, "How a Scientific Discovery is Made," *American Scientist -- Sigma Xi Scientific Research Society* 84 (1996) 364.

[2] *Ibid.*, p. 372.

Muller confessed: "Kepler's deep commitment to the guidance of three-dimensional structures of high symmetry — the five Platonic solids — was for me... a symbol of... holiness."[3] And in noting that their study of the influences is itself scientifically "precise, testable and generally applicable," Holton and his colleagues praise cultures that exhibit what the pioneer of the modern sociology of science, R. K. Merton of Columbia University, has "memorably termed the Matthew Effect (a reference to the text: 'Unto every one that hath shall be given, and he shall have abundance...' Matthew 25:29)."[4]

Having hopefully weakened the assumption that theology has no warranted impact on either philosophies of science or science, Augustine's notions may be more readily considered. The consideration begins by distinguishing his allowance for true observations from Plato's rejection of them. The rejection reflects epistemic ties between Aristotle and Augustine but a gap between Augustine and Plato.

Plato and Necessary Truth

Whereas Augustine allowed for observational knowledge, Plato's Theory of Ideas was formulated precisely to account for knowledge despite a lack of true observation. In contrast to observable things that share in unchanging Ideas, there is only knowledge of Ideas. Plato's *Republic* 477d-478a, for instance, states that only "Scientific Knowledge"of Ideas has any veracity by virtue of being infallible. Infallibility is not attributed to observation. The latter apprehends either changing particulars or their shadows and, epistemologically, involves the unstable states of mind of either Belief or Opinion.

These states of mind contrast to Understanding that apprehends invisible geometric entities, say "$\pi r \phi / 180$"as a Circle, where "ϕ" equals 360 degrees, and to Knowledge that grasps Ideas such as Circle or Cube in all of their geometrico-philosophical aspects. Though partaking of an Idea, a physical circle or cube *per se* begets erroneous observation since,

---

3 *Ibid.*, p. 372.

4 *Ibid.*, p. 375.

actually, the figure is continually different things, if any thing at all. There can be no truth about any given physical thing if it "is not what it is" in virtue of a change that attaches to things which are observed. Accordingly, there is no observational knowledge.

By contrast, in addition to articulating observational knowledge in a discerning way, examined shortly, Augustine apparently held that Plato's "a priori knowledge" would only be infallible by being trivially true. On the one hand, Augustine had an advantage over Plato insofar as he was evidently cognizant of Aristotelian logic. This cognizance is clear by his distinction of "true propositions" from trivially true "rules of inference" and "inner rules of truth".[5]

On the other hand, his previously noted assertion that it is better to know true propositions than inference rules takes on a novel spin. The rules are only limitedly applicable to thought about God from "whom... everything is derived."[6] As everything else, the rules are derived from God who created and can end them: *If all things will pass away including knowledge, with the exception of Love as the Scripture says, is it not the case that a knowledge of the rules of reasoning may be temporal as well?*

Begetting humility for logicians, temporal as well as cognitive limitations of logic are clearly expressed when Augustine says that "If... reason discerns something eternal and immutable, reason should admit that it is inferior and that this is *its* God."[7] God is Creator. Without being inconsistent with his realism wherein how *is* reality is the truth-condition for empirical "truth", Augustine was well known for his reluctance to render reasonable the Creator on the basis of flesh-and-blood reasoning and experience. Thus although he asserts that "immutable rules [were] not instituted by men but were discovered"— where discovered "rules" refer to the rules

5 For example, among other writings, see St. Augustine's *On Free Choice of the Will*, pp. 66-67.

6 St. Augustine, *On Christian Doctrine*, p. 73.

7 St. Augustine, *On Free Choice of the Will*, p. 49, emphasis added.

of inference,[8] there should be epistemic caution. Their supposed rationality is not to be associated surreptitiously with reality and they may hold only as long as does the existence of the world that is the truth-condition for worldly knowledge.

Intriguingly, this view of knowledge may have influenced Thomas in terms of an inapplicability of the rules of inference, viz. *modus ponens, modus tollens, hypothetical syllogism, disjunctive syllogism* and so forth, for angelic beings who have no need of them. Thus he says in the *Summa Theologica* (I, 79, 9) that whereas angels have an immediate grasp of perfect intelligible "truth simply and without mental discussion,... man arrives at knowledge by advancing from one thing to another". With Augustine's caveat that not all truth is perfectly intelligible, Thomas' insight is reminiscent of his position that mutable human reason struggles to arrive at truth, "sometimes reaching it and sometimes not." [9]

Taken with Augustine's recognition of realities enigmatic to reason altogether, as soon expanded on, questions also ensue about limitations of the laws of thought such as the Principle of Non-contradiction. And echoed again by Thomas who says "it is possible that anything... be created by God" in reply to the objection that God could not make "affirmation and negation... both true at the same time" (I, 45, 2), Augustine underscores that there cannnot be appeal to logically necessary or empirical truth for rendering reasonable the historicity and nonhistoricity of the Son of God. The same holds for the Trinity who, being a *mystery*, is both a unity and plurality.

Augustine is clear about the rational pretensions of philosophers: Their "wrangling" about an irrational Trinity begs for articulation of an enigmatic one in themselves — epistemic charity should begin at "home". They experience an inner trinity:

> far different from the Trinity, but I suggest them as a subject for mental exercise... In these three - being, knowledge, and will—

[8] St. Augustine, *On Christian Doctrine*, p. 72.

[9] St. Augustine, *On Free Choice of the Will*, p. 49.

> there is one *inseparable life*... although they are *[conceptually] distinct* from one another...[10]

Rejecting the Trinity because it is does not conform to our rational conceptions proceeds *pari passu* with rejecting an incontrovertible triune experience of ourselves — not that this has prevented philosophers from absurdly reducing themselves to material bodies or to minds or to wills, to which the history of philosophy abundantly testifies!

Truth in Theology and Science

The foregoing point of limitedly understanding God concerns faith not being entirely reasonable and limitations of cognitive truth. Logically necessary and empirical truth are not ascribed to physical *reality* but rather to a proposition per se or to a statement about reality. We no more seek a "true star" than demand that a star abide by the Principle of Noncontradiction. Rather, beyond noncontradictory propositions, we seek for them to be true. Who can assert *a priori*, prior to experience in Platonic fashion, that it is a rational necessity that reality must be a given way ? The assertion is itself irrational since we do not begin with reason but rather with how *is* reality as the condition for observation-scientific truth.

At the same time, to say there is nothing self-contradictory in asserting that reality need not abide by the Principle of Noncontradiction is to say that God need not abide by it either. If God is the Reality who created empirical reality and the latter is the condition for secular truth-claims, then the point that revealed truth cannot be made reasonable by a scientific rationality can scarcely be disputed. Augustine refers to God's revelation "I AM THAT I AM" for viewing God as the ultimate Reality on which empirical reality depends. The dependence begets a rationality of scientific reasoning, based on reality, as opposed to a rationality of scientific truth per se. And while there are not inferences from this truth to a revealed one, the former does not conflict with the latter.

---

10 St. Augustine, *Confessions*, p. 318, emphasis added.

From the Trinity *to* Jesus' historicity and nonhistoricity *to* unified and separate elements of our existence *to* persons being both free and not free (involuntary causes), the epistemology brings to mind a host of scientific issues. The issues range from an Underdetermination-of-Theory-by-Data (UTD) thesis in which logically inconsistent theories may be empirically equivalent to De Broglie's particle-wave equation in quantum physics wherein light is interpreted as both a wave and non-wave on the basis of observation. But while a statement "Light is a wave" may be true and false in apparent violation of the Non-contradiction Principle, despite different experimental contexts,[11] *the contradiction is not the irrational one of Sophism that starts with opined truths of possibly contradictory truth-claims for determining how is reality.*

Modal Scientific Reasoning

With recognition that Aristotle's modal logic is subject to *de dicto* interpretations where "necessity" pertains to propositions as a whole and *de re* interpretations where it is attributed to predicates of things, addressed shortly, there are three points. First, Augustine may accept an *epistemic* interpretation of Aristotle's Principle of Noncontradiction without accepting an *ontological* one. Whereas the latter specifies that something cannot both "be and not be", the epistemic one holds that a statement which is both true and false is inconceivable. But to admit of inconceivability is not to admit of an impossi-bility about reality.[12]

11 Notwithstanding the influence of Sir Karl Popper, noted below, see physicist Saul Youssef's scientific realism. There is no appeal to contradictory theoretical properties or constructs apart from evidence: The two-slit experiment *still* means prima facie that light is "both a particle and a wave", in "Is Quantum Mechanics An Exotic Probability Theory?," *Fundamental Problems in Quantum Theory* (NY: New York Academy of Sciences, 1995), p. 904.

12 When Sir Karl Popper laments our "anti-rationalist atmosphere" in reference to relativistic De Broglie-like formulae, he evidently either *(i)* confuses the formulae with an irrational Greek Sophism of starting with "truth" for determining reality or *(ii)* irrationally demands *a priori* that reality conform to the Noncontradiction Principle. See his *Quantum Theory & the Schism in Physics* (NJ: Rowman & Littlefield, 1982), p. 142.

We do not demand of reality that it abide by the Principle but rather that do statements. Without denying the Principle's logical necessity, it may be interpreted in terms of our limited cognitive powers by a standard modal symbolization:

$$\Box \sim(p \wedge \sim p),$$

where the symbol "$\Box$" designates "Necessarily $\sim(p \wedge \sim p)$" and the principle's logical necessity is not denoted but rather what is necessary for conceivability. But the Principle's applicability to reality may be represented:

$$\sim\Box \sim(p \wedge \sim p),$$

where "$\sim\Box$" may be read "It is not necessarily the case that".

Second, Augustine's references to the Trinity and tripartite conception of ourselves, for suggesting limitations of reason, suggest a threefold interpretation of the Principle of Noncontradiction in which it has *(i)* a logically necessary truth since its denial is self-contradictory, *(ii)* a modally necessary truth by virtue of indicating what is necessary for conceivability but not for what obtains ontologically and, given the latter, *(iii)* a limited ontological applicability because it may not always reflect given domains of reality. Again, there is nothing self-contradictory in saying that reality need not abide by the Principle of Noncontradiction.

Implications seem apparent for certain religious miracles said to be impossible, but we reiterate that these results are not those of a sophistic relativism. They reflect a realism of "truth as conformity of thought to reality" *(adequato rei et intellectus)* and, intriguingly, may indicate overlapping epistemic insights in Western and Eastern traditions without excluding such differences as Eastern distinctions of reality from contradictory conceptual constructions.[13]

---

[13] *E.g.*, this distinction is made in a dialectical logic (*prasanga*) of Nagarjuna's Madhyamika system.

Third, the Principle's modal necessity may serve as a regulatory principle. We may try to render noncontradictory or consistent conceptual-linguistic constructions for purposes of simplicity or the like and, that having been done or not done, be sure they have empirical truth which rests indirectly on observation. Scientists may argue that superseding theories in the history of science are not derived directly from observational truth. But a rational notion they are approximately true, in contrast to a rationality of truth per se, rests on predictions or manipulations whose observational descriptions are true. That this is not at odds with Augustine is clear by his remark that by "description... of the position and motion of the stars, predictions concerning the future [are] made according to rule which are... fixed by calculation."[14]

Taken with his realism, we may ask how the calculations could be systematically successful apart from phenomena being truly described by theories?

Having enumerated these important points, we note that some proponents of both *de dicto* and *de re* interpretations of Aristotle's modal logic may claim that the foregoing goes far beyond what Aristotle says. But besides the disagreement about what he says, restrictions by analogy to what Plato *says* would prevent historians of philosophy from attributing Platonism to Augustine's thought; not to mention the early Aristotle's![15] Furthermore, cases may be made for both interpretations.

When Aristotle speaks of forming syllogisms by relationships between *As, Es,* and *Gs* and adds that the "method is the same whether the relation is necessary or possible," in the *Prior Analytics* (45b), a *de re* interpretation seems evident. And when he says "It is necessary indeed, if animal follows man,

---

14 St. Augustine, *On Christian Doctrine*, p. 66.

15 In correspondence from Professor Emeritus R. Connell whose scholarship is well regarded but with whom one may reasonably disagree, as Professor Marcus indicates below, there was sole allowance for an apparent *de re* interpretation: Aristotle's modal logic is limited to "connections between subjects and predicates... expressed as modes over and above the truth or falsity that defines propositions... Nothing more than that appears in Aristotle's work on propositions."

that it should follow all these also," a *de dicto* view seems equally apparent in which "necessity" may be attributed to that statement as a whole. Such a statement is expressed, for instance, in Ruth Barcan Marcus' analysis of an Aristotelian essentialism in her *Modalities* (1993).[16]

Finally, objections in either case that Aristotle viewed such terms as "necessity" and "impossibility" apart from truth not only again collapses distinctions between his thought and developments of it but ignores various assertions;[17] as when, after referring to "modes of predication" (46a), Aristotle speaks of a "pursuit of truth starting from premises in which the arrangement of the terms is in accordance with truth"; and as when, in relating attributes that are essential and necessary to "necessary premises", in the *Posterior Analytics* (73a), he states that "if it be true to say 'this is a man', 'this is an animal' is also true". Prima facie, the latter may be expressed as an impossibility that "this is an animal" is false when "this is a man" is true as well as "Necessarily if this is a man *(M)*, this is an animal *(A)*":

$$\Box(M \longrightarrow A)$$

And prima facie it is not an empirical ("accidental") or logically necessary truth but rather a modally necessary one with an epistemological strength, regarding "truth" per se, that lies between them.

---

16 See Ruth Barcan Marcus, *Modalities: Philosophical Essays* (NY: Oxford University Press, 1993), p. 69. In discussing Aristotle's modalities in terms of an evident *de dicto* interpretation, "necessity" is held to pertain to a causal conditional as a whole:

$$R(S) \longrightarrow_c D(S), \text{ where } S \longrightarrow_c P =_{df} \Box(S \longrightarrow P).$$

17 *Cf.* C. Yang's "Necessary Truths, Essentialism and the Principle of Revisability," *Philosophical Review* 18 (1995) 187-211. With the caveat that modally necessary truths remain obscure in modern analyses of metaphysics, it is argued that adding "It is necessary that" to a statement *P* which describes something that actually happened may be as intuitively sound and meaningful as *P* per se as well as bear a definite truth-value.

One may object that this "strength" is odd: Since the empirical is an *epistemic* category and logical necessity an *alethic* one, a nomic necessity is begged for without a scale between them.[18] But imagine a logic teacher's criticism of students who might suppose that a two-conjunct proposition P_1, with one true conjunct, borders P_2 with two true conjuncts and P_0 with two false ones. Bearing on everyday decisions and even theory choice in science, there may be a sense in which P_0 and P_1 are not "equally false" and in which P_1 has an epistemic strength between the other two despite different ordinary-language and sentential-logic categories.

Why do the categories in question exclude "Necessarily if this is a man..." having the given strength between "This is a man or not" and "It is not" even if the latter is empirical? And an objection that the empirical is an *epistemic* category raises a good question: Following Kant's *Critique* with a modal category — despite its not penetrating modern logic, why may not logic admit of a modal truth between extremes of "truth" known analytically and a posteriori?

In virtue of the latter being synthetic, may there be an ignored mode of "synthetic a priori" metaphysico-modalities the truth of which excludes *a prior* some empirical claims, *e.g.* "Φ is a man and not an animal"? And in regard to necessary nomic laws that imply counterfactuals, *e.g.*, all Φs are Ψs and if a thing *were* a Φ it would be a Ψ — that have been well perused, a reason for this discussion is to probe, precisely in a novel way, how Augustine's idea of "necessity" may be related to a modern issue of theoretical truth.[19]

For purposes of either ascribing "truth" to theories or speaking of scientific "knowledge" *per se*, theories may be

18 I am indebted to Professor Del Ratzsch for this possible objection. "Nomic necessities" often refer to laws of nature with counterfactuals (noted below).

19 Cf. R. Boyd's "Realism, Approximate Truth, and Method," *Philosophy of Science*, Ed. D. Papineau (NY: Oxford University Press, 1996), pp. 215-255. Interestingly, Boyd suggests, without referring to nomic necessities, that laws and theories could not "yield predictive success unless... by and large true [Ed]."

expressed as compound propositions whose conjuncts are laws, *e.g.* the three laws of Newton as

$$(N_1 \wedge N_2 \wedge N_3)$$

and their interpretations such as free fall $mg = md^2/sdt^2$ for the second law $f = ma$ (N_2). At the same time, the truth of laws and theories cannot be verified, by a material-implication reasoning, because it is logically possible for them to be false when their predictions are true. Yet, may this reasoning at times disregard common sense? Augustine's defense of a *sensus communis* of persons is tied to his idea that what is *logically possible* may be *impossible* in terms of experienced things. There are "two kinds of falsehood, one which involves things that are not possible, and the other... that are possible but nevertheless do not exist".[20]

Though one falsity is that it rained on a given day which "might be true,"[21] another is the denial of a rule or law. Bringing to mind Aristotle's essentialism when it is coupled to scientific principles being "judged by their [perceptible] results" in the *De Caelo* (306a), the law is successful in explicating and predicting how persons walk wherein "the rear foot is not raised until the first foot is put down" which when denied prompts Augustine to say "Nor is it *possible* to walk in any other way."[22]

Keeping in mind that a biophyical human nature is rooted in Nature, the logic may be applicable to the natural sciences: "It is impossible for their laws to be entirely false when it is true they are perceptibly and systematically successful in given domains": "Necessarily if the laws have such success, they are approximately true". Unless they truly reflect how is Nature, how could they successfully explicate and predict its behavior?

On the one hand, it holds prima facie that "Necessarily if *X were* a person walking, *X's* rear foot..." and "Necessarily if *X were* a man, *X* would be an animal" because they reflect

---

20 St. Augustine, *On Christian Doctrine*, p. 71.

21 *Ibid.*, p. 71.

22 *Ibid.*, p. 72. Emphasis added.

primary morphological natures of things. Evidently, however, the necessity does not hold for modern formalized laws which do not reflect such prima-facie natures. Further, it is modally fallacious to reason "Necessarily if the laws have such success, they are approximately true" and "They have the success" so "*Necessarily* they are approximately true".

On the other hand, it seems equally evident that the conclusion "They are approximately true" is still true. And a modal necessity may in some sense hold, as noted later, for a historical sequence of well tested "domain-nested" laws when viewed collectively.

Paradigm doubt about the laws surrounds Newton's first law. But even if it is not descriptively true per se, it may have a truth by way of the second law that may be regarded as a well-tested theoretical description of properties and relationships: One might object that bodies do not accord with the first law "All bodies unacted on by force have a constant velocity...". Yet, in defining a set of reference frames where the mechanics hold, the first law may be included in the second as a special case where bodies will not accelerate without force (in terms of a simple vector equation):

$$\sum \mathrm{F} = ma$$

Consequently, though Augustine's rule or law on walking is straightforwardly descriptive as opposed to the Newtonian law, the latter's linkage to a systematic predictive success may render unreasonable a denial of its truth. How can the law be entirely false, in reflecting how is a domain of reality, when it is integral to systematic success?

This question brings to mind recent insights in the philosophy of science. For example, Nancy Cartwright and Ian Hacking lend their thought to the notion that truth is an *existential ingredient* in causal explanations. Hacking argues that experimentalists do not believe in electrons since they "save the phenomenon" but rather because they use them to manipulate and create new phenomena. And Cartwright augments his argument by adding that, while truth is an "external characteristic" of laws or theories, physicists *must* be committed to the real existence of a cause such as an electron if they accept a given causal account.[23]

That is, it is impossible not to be committed to the real existence of the cause when accepting a given causal account. Recasting the impossibility as "Necessarily if there is acceptance of a given causal account, there is commitment to the real existence of the cause" seems to proceed *pari passu* with "Necessarily if there is commitment to the real existence of the cause, there is acceptance of the approximate truth of a theory in terms of which the cause is understood".

What sense does it make to speak of a really existing cause, say an electron, apart from a successful theory in terms of which it is understood? And the understanding, from J. J. Thomson to Niels Bohr to a present-day theory of the electron, accords with a realism in which the theoretical entity does not change but rather theories with increasing approximate truth.

That Augustine was quite familiar with these sorts of conditionals and that his thought may be coordinated with conditionals about our nature, is indicated by his recognition of a material implication in which there is a logical possibility that true "consequents may be derived from false antecedents" as well as by his modally relevant assertion that when:[24]

> true and certain propositions [*e.g.*, "persons walk"] are joined by valid inferences to propositions we are not sure about [*e.g.*, "the rear foot is not raised until the first foot is put down"], the latter, also, necessarily become certain.

That is, a modern view of conditionals, in terms of material implication, specifies *inter alia* the logical possibility that consequents may be either false or true when antecedents are true. But an elementary idea of a modal conditional $p \longrightarrow q$ specifies the impossibility of q being false when p is true: p entails q *if and only if* $p \longrightarrow q$ is necessarily true. And in precisely this manner, Augustine's reasoning falls within the domain of modal logic when he notes that it is impossible for

23 I. Hacking's "Experimentation and Scientific Realism" and N. Cartwright's "When Explanation leads to inference," *Phil. Topics* 13 (1982) 71-88, 111-121.

24 St. Augustine, *On Christian Doctrine*, pp. 69, 72.

some consequents to be false when the antecedents are true. With his recognition of logically possible truths which nonetheless involve things that "are not possible" and an ancient-medieval tradition that grounds human nature in Nature, that included Aristotelian modalities, are there not clear implications for the natural sciences?

One might say that this understanding of Augustine is *at least* as defensible as glib dismissals of his thought as an antiscientific Christian Platonism, evidence for which is enumerated shortly. We now note the relevance of his reasoning to a "lived" modal notion concerning the falsity of laws in everyday domains of kinematics, dynamics, and statics.

That the falsity of the laws is an impossibility when they are systematically successful even to modern logicians, even if unacknowledged seems evident by their trusting their lives to an engineering physics of buildings whose imminent collapse is generally disregarded. Apart from considering such things as earth-quakes or material fatigue, logicians would think someone strange who expressed alarm that the buildings might suddenly fall because of a "logical possibility" that well-established laws no longer obtain in the sense of reflecting how is a physical reality that is observed.

Observation and interpretation

This brings us back to Plato. For given there is no veritable observation, there would be no truth for empirically corroborating the substantive scientific knowledge to which Plato does himself appeal. "Believers", says Augustine, trust reports "of their bodily senses which subserve the intelligence. If they are at times deceived, they are at least better off than those who maintain that the senses can never be trusted."[25]

What cannot be trusted scientifically is a surreptitious everyday judgment that, say, an oar is really bent in the water. The judgment is not implied by the observation per se. An everyday judgment about an observation may need to be replaced by a theoretical judgment of science. The function of

25 St. Augustine, *City of God*, p. 466. Compare this to Descartes' *Meditations*.

science is to explain *inter alia* why the oar appears bent, which is how it should appear, and to predict the appearance.

W. H. Newton-Smith connects such appearances to current science. An appearance of white stars is consistent with an astrophysical theory that specifies other colors since, exactly, the white appearance is explicated and predicted. There can be no prediction without theory and no corroborated theory without a general faith in observation reports:[26]

> while we cannot have absolute faith in any particular reports..., we are entitled to have general faith in the low-level *O*-[observation] reports... If such judgments were not by and large reliable, we should not be still here to make judgments.

Could anything be more Augustinian? While scientific theories do not yield knowledge of God, Augustine holds that a history of science may, from a sheer philosophical standpoint, be compatible with Scripture. In noting that Scripture is a "history of past things, an announcement of future things, and an explanation of present things,"[27] he suggests that scientific theories which are inconsistent with Scripture may evolve into ones in the furture that are consistent.

This suggestion brings to mind the biting comments of physicist W. Schaaffs. He states that many modern physicists and theologians alike relegated a Scriptural "beginning of the world" to pure mythology before an expanding universe implied the beginning "in a very small place".[28]

Regardless of its being a submicroscopic black hole with maximum entropy addressed by both quantum and relativistic physics, Max Born once observed that an association of religion with the world's beginning led many scientists to seek to avoid it with a static concept. Further, God's creation of light before our sun was equally concluded to show that the relevant passages in Genesis were either senseless or pure mythology. Now, however, light is construed as a primary electrical and

---

26 W. Newton-Smith, *The Rationality of Science* (London: Routledge & Kegan Paul, International Library of Philosophy, 1980), p. 28.

27 St. Augustine, *On Christian Doctrine*, p. 88.

28 W. Schaaffs, *Theology, Physics, and Miracles*, Tr. R.L. Renfield (Washington DC: Canon Press, 1974), See pp. 35, 41.

atomic effect, and the formation of a sun or fixed star is often interpreted as a secondary macrocosmic effect.

One would commit the fallacy of "missing the point" (*ignoratio elenchi*) in objecting that these developments may themselves be superseded by theories inconsistent with Scripture. The point is simply that most modern philosophers of science and scientists view science, historically, as reflecting a verisimilitude of increasingly true or true-like theories. Such theories may seem to cut two ways because any theory that seems consistent with Scripture, at a given historical time, may be superseded by truer theories which may be inconsistent. However, the verisimilitude indicates a general direction of theoretical developments which is certainly not adverse to a traditional hermeneutics of Scripture.

That is, while the objection may ensue that a development of future theories may yield the adversity, it ignores that the adversity supposes a development of theoretical truth that, being at odds with an increasing millennial truth at hand, is at odds prima facie with the verisimilitude to which the objection appeals. Cornell physicists F. Rohrlich and L. Hardin bear on exactly this point. In a modally relevant manner, they refer to a historical growth of domain-nested theories. The theories are typically so well tested that it is *more* than unreasonable, and not merely unreasonable, to suppose that the theories will turn out to be entirely false unless Nature itself radically changed.[29]

---

29 F. Rohrlich and L. Hardin, "Established Theories," *Philosophy of Science* 50 (1983) 603-617. Modal reasoning, related to a historical growth of domain-nested theories, is elaborated on in the discussion of Thomas.

Chapter 3

Critique of Critical Thought

After initially expanding on Augustine's notion of "truth", the puzzling question is considered of why an Aristotelian epistemology may have been ignored. Subsequently, there is an overview of the epistemology and an examination of how it challenges a central core of Humean-Kantian "critical thought". Since the latter has shaped a central core of the philosophy of science and Augustine is in general accord with its realism, his thought may strengthen that view.

Thus, with regard to "truth", it is initially noted that Augustine's view on universal scientific ideas seems viable. The notion that theoretical properties, processes, and relationships are exactly or inexactly embodied by observable things, in either Aristotelian or modern physics, renders intelligible predictions. For example, the intelligibility of predicting appearances of bent oars in water seems to rely, at least in part, on causal relationships of water and light. Light permits veritable observations of phenomena, necessary for formulating universal scientific laws or theories, as well as testing them by predictions. And this holds despite an Aristotelian physics that eschewed testing.[1]

---

[1] An Aristotelian concern to not alter Nature by testing was abandoned when science required alternative assumptions. But his concern was not naive and, over two millennia later, echoed in quantum physics. Thus although the concern is irrelevant to instrument-aided obsevation of, say, orbiting planets, it is relevant to microphysical phenomena disrupted in experimental setups. Cf. D. Lindberg, *The Beginnings of Western Science* (Chicago, Ill: The University of Chicago Press, 1993), p. 53.

Importantly, testable knowledge may supersede everyday truth-claims when they pertain to how, approximately, physical reality *is*. At the same time, everyday claims cannot be dismissed cavalierly. In drawing apparent implications of Augustine's thinking, for instance, the claim that the sun is rising may mean that it is time to get up. The claim is not a false scientific one that the sun really rises but rather a true everyday one.

Whereas the everyday claim may serve a certain purpose that makes it no less true on that account than a different purpose of science, scientific claims may involve the judgment that one body moves relative to another. Observational descriptions of the relative movement might be coordinated with either Ptolemaic or heliocentric theories that beg for testing. Nonetheless, scientific and everyday claims may stem from the same appearance.

We can as much certify that it is time to rise by an appearance-laden judgment that the "sun rises" as we can that an oar is not really bent by virtue of an optical science that explicates and predicts appearances based on the senses. Unless "what we perceive... passes beyond the inner sense," notes Augustine, "we cannot arrive at knowledge (*scientia*)".[2] This understanding of science with the modally suggestive term "cannot", whereby our senses not only lead to mundane truths but to scientifically explained appearances, is not naive.

Augustine's understanding begets more than a superficial connection to Aristotelian modalities in which one might say that it is *impossible* to have knowledge (*scientia*) when one does not pass beyond one's bodily senses : Necessarily if one does not go beyond the senses, one does not obtain knowledge (*scientia*). The knowledge involves Aristotelian forms as well as causes ignored by historians of philosophy.

Aristotelianism Ignored by Scholars!

A close connection to Aristotle is evidenced by Augustine's stress on "outer" and "inner" forms. In the *City of God,* he says that "Every material body has an outer form shaped by a potter... or

2 St. Augustine, *On Free Choice of the Will*, p. 42.

other artisan who can paint or fashion even forms that look like the shapes of animals."[3] He adds that "there is also an inner form which is not a shape but a shaper, with an efficient causality deriving from the secret and hidden determinations of some living and intelligent nature which can shape... the inner souls of living things."[4]

Reference to things that have inner forms with efficient causality is inextricably related to a pioneering incorporation of Aristotle's thought into Christian philosophy. Despite Augustine's relative indifference to proofs of God, his philosophy had a prima-facie influence on Thomas' proofs by way of motion and efficient cause. Thus although a First Mover and Cause have long been admitted to connect Aquinas to Aristotle, an epistemic connection of Aristotle to Augustine has been neglected by many historians of philosophy — as well written and insightful as are their works.

For example, S. E. Stumpf says, with no reference to Aristotle, that "Neo-platonism had finally made Christianity reasonable to him [Augustine]".[5] W. T. Jones, while properly credited for indicating how philosophical history was influenced by the history of science,[6] merely asserts that Augustine "found in Neo-platonism a notion of the deity as a creative force... rather than as a crudely anthropocentric architect or a handicraft worker".[7] And E. M. Albert, T. Denise, and S. Peterfreund say, without any reference to Aristotle, that Augustine "turned... to Neoplatonism".[8]

One may reasonably suppose that a largely abandoned medieval tradition, coupled to impatience with scholarly

---

3 St. Augustine, *City of God*, p. 264. An "outer form" implies, prima facie, an "inner form".

4 *Ibid.*, p. 264. Emphasis added.

5 S. E. Stumpf, "St. Augustine's Christian Philosophy," p. 135.

6 F. Suppe, Ed. with Critical Introduction and Afterword, *The Structure of Scientific Theories* (Ill: University of Illinois Press, 1977), p. 717, fn. #258.

7 W.T. Jones, *A History of Western Philosophy: The Medieval Mind, Volume II* (NY: Harcourt Brace Jovanovich, 1969), p. 79.

8 E. M. Albert et al, *Great Traditions in Ethics* (NY: D. Van Norstrand Company, 1988), p. 84.

nuances and even a concern for teaching simplicity, led to a stratagem that Platonism makes sense of Augustine's mere concern for religio-moral matters. By analogy to Socrates' simple concern for Virtue in contrast to Plato's complex Theory of Ideas, Augustine's thought allegedly contrasts to Aristotle's formal logic and physics.

The furthest thing from a rigour of physics is suggested by the titles of Augustine's *On Christian Doctrine, Confessions, City of God,* and *On Free Choice of the Will.* Free choice (*liberum arbitrium*) of the will, in particular, may seem to conflict with a deterministic science and to underlie his anti-scientific stress on our responsibility for salvation.

And a Platonic contrast of God's illuminated immutable realm, relevant to salvation, to an impenetrable darkness of the human one may have entrenched the interpretation. As Professor Kaufmann suggests, Augustine's stress on our alienated existential condition makes him a forerunner to the modern existentialists. With their mood of despair and revolt against rational systems, they are in stark contrast to systems of Aristotelian science as well as to a scientific optimism of the Enlightenment.[9]

Finally, with sensitivity to both sides of a religious revolution, which in some ways facilitated the Enlightenment, it needs to be said that the Protestant Reformation may have exacerbated ignorance of Augustine. In breaking with a sacred tradition of religious and metaphysical developments, from Augustine to Thomas, the Reformation weakened any left over medieval metaphysics as well as strengthened ties of philosophy to science. Science came to be viewed in many quarters, positivistically, as altering the nature of philosophy and as superseding theology.

Tellingly, the reformed theology was celebrated by Marx because it "turned priests into laymen and laymen into

---

9 W. Kaufmann, *Existentialism From Dostoevsky to Sartre* (NY: NAL Penguin Inc., 1975), p. 13. For criticism of Kaufmann's sympathy with an Enlightenment of the Greek Sophists to whom Nietzsche is compared, see R. Trundle's "Cold War Ideology," *Res Publica*: *Leuven Politologisch Instituut* 37 (1996) 61-84.

priests."[10] Putting religion at the mercy of either democratic or egalitarian social-political trends, these new "lay priests" provided a secular milieu for the revolutionary philosopher. Where the philosopher's "speculation ends," said Marx, "there real, positive science begins."[11] Since modern science began by rejecting the four causes, modern philosophers assumed that they should reject a causal metaphysics — quintessentially a metaphysics of Thomism influenced by Aristotle.

Aristotle thought was conceptually infused, from start to finish with Thomistic scholasticism. The latter was largely ignored in modern English universities, such as reformed Oxford or Cambridge, or taught in American colleges with secular or Protestant affiliations. Hence, Augustine's Platonism made his theology less of a slippery slope to Thomistic Catholicism and less threatening to a view of science as a paradigm "successful epistemic enterprise...indicative of the nature of knowledge."[12] His supposed antagonism to the knowledge encouraged philosophers to make a clean break with medieval theology in order to share in a "progressive" scientific revolution by way, among other things, of Kant's "Copernican Revolution".

Unhappily, despite Kant's mental category of modality, his Revolution disregarded that modalities might comprise an epistemological boundary between *analytic* necessities and *a posteriori* contingencies. Related to "truth" and not to a truth-valueless metaphysics, the ignored modalities complemented reformed works such as Luther's *De servo arbitrio* (1525) and Calvin's *Christianae religionis instituto* (1536). The latter neglected any "worldly" theological modalities in favor of earlier Platonic-Augustinian doctrines. These included free will and divine grace that, unrelated to phenomena, seemed to put faith in its proper noumenal place of another "world". Such developments resulted in political "lay states," notes historian

10 K. Marx, "Towards a Critique of Hegel's Philosophy of Right," *Karl Marx: Selected Writings*, Ed. by D. Mclellan (NY: Oxford University Press, 1987), p. 69.

11 K. Marx, "The German Ideology," *Karl Marx: Selected Writings*, p. 165.

12 F. Suppe, "Afterword," *The Structure of Scientific Theories*, p. 716.

S. Schapiro, that were "neutral to all faiths... restricting each to the purely spiritual realm."[13]

Harvard honors graduate John Cort notes that this other "realm" was not merely ignored in terms of Augustine. Not one "course in philosophy dealt seriously with Thomism. The introductory course... jumped from Aristotle to Descartes... nearly two thousand years during which, according to the department, there was no philosophy worth mentioning."[14] And written by Yale graduate W. Wilson as a refresher book for college graduates, *An Incomplete Education* (1987) echoes the neglect's impact on the educated public. The book makes no reference to Aristotle in discussing Augustine and it says of Thomas that he is "a lesson in thoroughness;... unless you're a Catholic priest, take him with a grain of salt."[15]

Unmistakable Aristotelian Causality

Having seen why Thomas as well as Augustine's Aristotelianism were ignored by philosophers in the positivist tradition, an import for modern science is noted after further noting a cosmic Aristotelianism. Also, there will be consideration of possible challenges by some scholars of medieval philosophy.

Thus, Augustine says "God is the Cause... that *makes* but is not made...". And he adds that "Material causes... are not... efficient causes, for their power is limited to what the wills of spirits work through them".[16] These causes are not even in Plato's vocabulary but only in Aristotle's. Coupled with the quotes below, they bear on the notion that all creatures speak in a mystical code (*secreto quodam loquendi modo*) to intimate "the Trinity every time we ask: Who made it? [God the Father

---

13 S. Schapiro, "Anticlericalism," *Ency. Brit.* (Chicago, Ill: W. Benton, 1969), p. 63.

14 J. C. Cort, "A Bizarre Conversation," *The New Catholics*, Ed. by D. O'Neil (NY: Crossroad Publishing Co., 1987), p. 5. He adds that "I once told the wife of a Harvard professor that I had been converted to Catholicism while an undergraduate there, and her reaction was, "How positively bizarre!"

15 W. Wilson, *An Incomplete Education* (NY: Ballantine Books, 1987), pp. 308-309.

16 St. Augustine, *City of God*, p. 108. See p. 233 for quote below.

as Efficient Cause] How? [through the Son as Formal Cause whose Truth is intimated by genus and species] and Why? [since the Holy Spirit as Final Cause purposes Love]."

Despite Thomas' *Summa Theologica* (I, 45, 7) that suggests a similar insight, a notion of Aristotle's pagan God as Form of the world's substance may fuel objections by some modern scholars:[17] Augustine refers to the Trinity and not to the Causes since, given that God is not a material cause, the causes would include the pagan Formal one. Yet questions ensue of whether rigid interpretations of Thomas, excluding any sense of divine Form, evoke "Thomistic" interpretations of Augustine. Given a neglect of his Aristotelianism, it seems prudent to be cautious about such interpretations and *a fortiori* the objections.

God is not a Material Cause since He created the world *ex nihilo.* And while God is not merely an immanent Formal Cause in terms of an Aristotelian *Nous,* He is present to the world as an *external* Efficient Cause and an immanent Intelligence from which derive inner forms. Thus although a divine Efficient Cause is a presence giving "being to... genus and species,"[18] Augustine says both that "there is also an *inner* form... with an efficient causality *deriving* from... some living and *intelligent nature* which can shape not merely the outer forms of physical bodies but the inner souls of living things" and that:

> It was this by this same divine creative force... taking on their *form* with *no extrinsic cause* but by the *intrinsic power* of the Creator, who said: 'Do not I *fill* heaven and earth?' and whose wisdom 'reacheth from end to end mightily, and *ordereth* all things sweetly.'

Since efficient causes are derived from an intelligent nature, there is both a voluntary causality underlying our self-conscious intelligence and a divine ordering intelligence in Nature. This brings to mind modern ideas of a "self-organizing" biophysical intelligence that is augmented by Augustine's thesis of *rationes*

---

17 Though I am beholden to Professor Emeritus Norris Clarke, S.J., for clarification of these Causes, a divine formal cause is held by him to result in pantheism (to which I have responded). Here, a possible reason for the rejection of a formal cause is further considered.

18 St. Augustine, *City of God*, p. 266. See pp. 264 and 265 respectively for quotes below.

seminales. It posits seminal principles in evolving organisms and suggests, beyond Aristotle, a consistency of evolution and creation. Aristotle denied creation on a dogmatic assumption that only unities of form and matter were real: Ostensibly, there would have to be time before creation which could not obtain because time is the before and after of motion and there would be nothing in motion. Nevertheless, in suggesting a gradual evolution in his *History of Animals* (588b), Aristotle indicates an evident influence on Augustine. In sum, Darwin's assertion that the pioneers of modern anatomy were "schoolboys to old Aristotle" suggests that Augustine's work has an ignored sophistication as well.[19]

Causality and Metaphysics

Augustine's sophistication, in terms of his thesis of free will, bears on tenable epistemologies. They should permit a common-sense view of ourselves as both voluntary and physical causal agencies. Modern philosophers and scientists came to accept only empirical truth that, assumed by scientific inquiry, was applied exhaustively to human behavior and found expression in the causal principle "All events are caused". Appeal to it led to a skepticism in which scientific truth was undercut since the principle was unverifiable but presupposed. It was not properly viewed as a part of "things after physics" (*ta meta ta phusika*) related tenably to a broader metaphysics.

Whereas a modern principle reduced causes to natural phenomena, a medieval principle specified both natural and voluntary agencies. A voluntary agency is implicit in Kant's notion "All persons are free agents". The medieval principle wedded what was divorced in his metaphysics of freedom and causal determinism. What was complementary was not only taken to be contradictory but freedom, the "good" and God were reduced to mere possible realities (*noumena*).

One may object that, according to Kant, the "noumenal world is the really real world as it is in itself, known only by a Creative Mind (Nous), which we cannot know. We know only

---

19 See Sir David Ross, *Aristotle* (NY: Barnes & Noble, Inc., 1964), pp. 112, 128.

the phenomenal world as it appears within our consciousness, *except when* we are thinking morally and infer the existence of a real — noumenal free will in ourselves, an important breakthrough from the phenomenal to the real world for us."[20] But this objection is not merely countered by Pope John Paul II who, echoing a religio-philosophical vista of the ages, invites sober reflection on a Kant-induced exit from our actual moral experience.[21] Precisely, the objection fuels its own rebuttal.

To argue that a noumenal world of things-in-themselves is really real, but known only by a Nous that is itself a thing-in-itself, is to beg the question of its reality. Kant says he "found it necessary to deny *knowledge*... to make room for *faith*."[22] But the words "except when," in the objection, make *faith* superfluous if we can infer a really real noumenal freedom. For how can the freedom be inferred from moral thought unless the thought involves knowledge of a religio-moral reality? Insofar as it does, faith is unnecessary in virtue of knowledge. And insofar as Kant says there is knowledge of "visible acts... subject to the law of nature,"[23] a notion of natural behavior being freely chosen is self-contradictory.

The causal principle rendered contradictory "Some events are not caused" in terms of some natural behavior being freely chosen. But Augustine allows for freely chosen behavior in virtue, among other things, of our self-consciousness: To be conscious of thinking is, not to infer, but, to be immediately aware of being free to think or not to think — including a paradoxical thought that thoughts are causally determined. No reference is made, which brings to mind Plato as well as Kant, to another world.

---

20 I am idebted to Professor Emeritus Norris Clarke, S.J., for this objection.

21 Pope John Paul II states: "If God is a knowable object... He is such on the basis of man's experience both of the visible world and of his interior world. This is the point of departure for Immanuel Kant's study of ethical experience in which he abandons the old approach found in the writings of the Bible and of Saint Thomas Aquinas." See *Crossing the Threshold of Hope*, Tr. by V. Messori (NY: Alfred A. Knopf, 1994), p. 34.

22 I. Kant, *Critique of Pure Reason*, Tr. N. K. Smith (NY: St. Martin's Press, 1965), p. 29.

23 *Ibid.*, p. 28.

Nonetheless, there are the two worlds of "heaven and... earth" in which there is a Creator.[24] And we are like him since, in not being mere natural creatures, we have artistic and scientific creativity. There can be no creativity without a self-conscious freedom to think of what *is* as well as of what *is not* but might *be*. Causally determined matter can be no more real than the immaterial freedom and thought in virtue of which scientists creatively formulate laws. It is senseless to object that we cannot "see realities that are immaterial".[25]

Reservation about the reality of an immaterial free will in this world lessens attention to the will of a Creator. The *Confessions* warns of this tendency in science. While recognizing scientific contributions, Augustine laments that astronomers flatter themselves too much when claiming credit for their predictions: Though "they can predict an eclipse of the sun..., they cannot see that they themselves are already in... eclipse.[26] He suggests that they may suffer a philosophical eclipse because they "do not inquire how they come to possess the intelligence to make these researches".[27]

Augustine holds that scientists ignore an *immediate* inner awareness of freely exercising their intelligence when they examine *remote* external phenomena whose causal sequences afford predictions. Thus, they will tend to accept only what is based on their physical senses. This insight bears on Hume's skepticism since particular sensations cannot yield a universal causal principle to which "truth" is ascribed, much less yield ideas such as "freedom" and "intelligence". Even if a divine Intelligence was entertained, there would tend to be dogmatic dismissals of any theological revelation.

Modalities and Critical Thought

Kant largely disregarded Hume's skeptical distinction between sense impressions and an external reality which supposedly

---

24 See St. Augustine, *City of God*, p. 536.

25 *Ibid.*, p. 536.

26 St. Augustine, *Confessions*, p. 93.

27 *Ibid.*, p. 93. They also typically conflate physics and metaphysics.

caused the impressions. But he held with Hume that the causal principle "All events have causes" involved only physical reality and was not *known* a priori. Hume's criticism, and Kant's response to it, go to the heart of a modern "critical thought" that supposedly examined presuppositions held, dogmatically, to be true.

Kant agreed that the principle is presupposed by scientific inquiry. In being supposed *prior* to the inquiry, the latter does not establish the principle's empirical truth. Kant ostensibly resolved the question of "truth" by transposing the problem of the causal principle not being known *a priori* into the thesis of an "*a priori* cognition" which interprets reality causally.

Widely ignored is that Kant's thesis is a "psycho-logical" response to a logical dilemma about truth. The mind, not truth about reality, is addressed. And scientific truth is undermined, in any case, since it presupposes a truth-valueless principle assumed a priori. Further, the "a priori cognition" is either a *phenomenon* or *noumenon.*

If the cognition is a *noumenon,* it is no less an unknown possible reality than freedom. And if it is a material *phenomenon* that interprets particular experience, how can it yield universal ideas such as a causal principle in regard to "all events..."? Both a psychic "freedom from" the material realm and causal regularities in it, for predictions, are necessary for an intelligibility of scientific knowledge.

Augustine related knowledge to an immaterial psychic self-consciousness that was not reduced to Hume's "species of sensation" or to Kant's cognition. He appealed to our immediate awareness of free will and suggested we are aware of our freedom to reason or not to reason — including a self-refuting scientific reasoning that it is causally determined.

The issue of determinism and freedom is not resolved by a truthless metaphysics of phenomena and noumena that is itself *synthetic a priori.* In permitting only empirical (*synthetic*) and logically necessary (*analytic*) truth since metaphysical and metaphysico-modal sorts of reasoning were evidently confused, modern philosophers ignored an alternative truth. Modal truth ties freedom to an incontrovertible awareness of our reasoning. After noting that even a "beast *could not* open its eye... unless...

aware that it does not see," Augustine argues that by our greater self-awareness "all these things, and reason itself... are included in knowledge."[28]

He underscores an impossibility that we scientifically reason when unaware of it and that to admit of the awareness is to admit of freely willing to reason. Recast as a necessity, there is: Necessarily if we are not aware of our will to reason, there is no scientific reasoning. When Dr. Paul Tibbets argues for reducing awareness to physico-neural functions because cognition is explained, do not the functions explain his own argument as well as counter-arguments?[29]

Epistemologically, materialism is tied to determinism for scientific purposes of causally relating temporal states of physical systems. But a psychic self-conscious freedom from the systems is as modally necessary for cogent arguments in psychology, despite a prevalent reductionism,[30] as in biology that there be prior reproductive processes for a person's conception — a necessity suggested by Wittgenstein, noted later, to border logically necessary and empirical truth.

Finally, Kant's thesis of an "a priori cognition" was no more empirically or logically true than the causal principle. Still, the dilemma of having to appeal to the principle's *determinism* for intelligible scientific inquiry explains a studious disregard of, if not antagonism to, Augustine's stress on *freedom* and its relegation, along with faith, to a noumenal world.

This only possibly existing world, that is not an object of knowledge, portended a pernicious outcome of Kant's Copernican Revolution. In including freedom, a rational mind, God, and God's judgment of our choices, a noumenal world was juxtaposed to phenomena. A reality of the phenomenal world was not in

---

28 St. Augustine, *On Free Choice of the Will*, p. 45, emphasis added.

29 See Dr. Paul Tibbets' book review in "History, Philosophy, and Policy," *American Scientist - Sigma Xi Scientific Research Society* 84 (1996) 409-410.

30 Also see MacArthur Fellow and Stanford University professor of biological sciences R. M. Sapolsky's "On Human Nature," *The Sciences* 36 (1996) 16, in which he states: "Nothing brings home the *material* nature of the mind as... the study of the brain with a positron emission tomography (PET) scanner."

doubt and was subject to a scientific inquiry that presupposed the causal principle. Taken with physics as ideal knowledge, held by the Enlightenment, the principle's covert acceptance made moot any religio-moral experience. In order to accomodate the experience, two worlds were articulated which resulted in religious skepticism and rational theology.

Rational theology arose since theology ignored a richness of ordinary experience related to God and an intelligent mind. In accepting only sense experience, "mind" and "God" were mere regulative ideas for a coherent synthesis of sense experience. *Religious skepticism* emerged since a coherence of the experience relied precisely on an intelligent mind and God whose realities were in question. In sum, there was no traditional God who was both a transcendent Being and an immanent intelligence in phenomena. But questions ensue of whether a supposed impossibility for phenomena in this world *is* both possible and analogous to a possibility for God.

A possibility of one God in two worlds is ignored by modern philosophers who either reject Christianity or seek to make it "reasonable". The latter fail to address scientific developments consistent with an intelligence infused into Nature by a Creator. A created Nature, with the intelligence, renders plausible the evolution of human nature with conscious capacities of reason and free choice of will. A perennial relevance of this point is exemplified by Bertrand Russell. He overlooked degrees of a voluntary efficient causality when he stated naively that "unless we are willing to give free will to the protozoan we cannot give it to man."[31] And in regard to a point now considered, the scientific developments are in accord with a traditional God who is transcendent and immanent. This point bears on Sidney Hook. In denying a rationality of ambiguity, he held that "An immanent God cannot be plausibly conceived as a personal God... and ambiguity is introduced into traditional religious faith."[32]

---

31 B. Russell, *The Autobiography of Bertrand Russell* (Boston: Atlantic-Little, Brown, 1967), p. 59.

32 S. Hook, "Modern Knowledge and the Concept of God," *Critiques of God: Making the Case Against Belief in God*. Ed. by P. Angeles (NY: Promethius Books, 1977), p. 33.

Chapter 4

Thinking Matter: The Philosopher's God

Augustine's caution against ignoring an intelligence in things draws attention to a problem of more open-minded scientists. Even if they acknowledge an intelligence that either intimates or is called "God", the latter would tend to resemble the philosopher's "God". Paradigmatically, such an intellectual divinity would be akin to a rational Aristotelian Nous. Far from inducing the specter of a revealed transcendent God, Augustine suggests that such a "Mind" would evoke the demand that revelation be reasonable. This is exemplified by John Locke whose "thinking-matter" thesis held that matter is infused with divine reason. Professor John W. Yolton asserts that this thesis led Locke to hold that "the principles he needs for his religion are few and simple. He makes Christianity a *reasonable* religion."[1]

Certainly, a divine reason permeates creation insofar as God, as already noted and soon expanded on, manifests an "inner intelligence" that "fill[s the]... earth". But Augustine complements this view by another. Whereas Aristotle held that divine reason governs choice, an omniscient reason of God is hidden by His omnipotent will. The will per se, divine or

---

[1] John W. Yolton, *Locke and French Materialism* (Oxford: Oxford University Press, Clarendon, 1991), p. 156. See both praise and criticism in R. Trundle's invited review of this work, quoted by the Clarendon Press, in *The Modern Schoolman* 67 (1991) 75-8.

human, is not fathomable by either science or epistemology. That is, it cannot be an "object" of knowledge. This becomes clear in considerations of a consciousness, that cannot be reduced to a phenomenon as soon expanded on, with which free will is inextricably related. Ultimate metaphysical questions, say either why the world is one way rather than another or why there is a world, are related to a principle of cosmic "will". And they are questions beyond rational answers.

One need only ponder a physicist's recent assertion that a self-organizing universe and "laws of physics...could very well have been spontaneously generated after Planck time [after the 'big bang']", which evokes the ideas of "choice" and "purpose" (*telos*), to see how physics is not at odds with a teleological metaphysics.[2] At the same time, limits of metaphysical thought cast light on Augustine's assertion that "We call Him omnipotent because He does whatever He wills... and suffers nothing... He does not will to suffer."[3] These considerations beg for a reappraisal of Augustine's contributions to modal thought.

Self-Conscious Thought and a Modal Impossibility

Interestingly, Augustine's reference to an infinite power of God's will does, again, lead to thought-provoking insights on modern philosophy. One may note, for instance, that John Locke, in evident *inconsistency* with his own empiricism, suggested that there is nothing inconsistent in the notion that God could superadd to matter the power of thought.[4] Professor Yolton holds

---

2 See V.J. Stenger's "The Face of Chaos," *Free Inquiry* 13 (1993) 13-14. In having maximum entropy, the fluctuation of the submicroscopic black hole involved a *teleological*, but unpredictable, self-organization of matter evolving into self-organizing biological systems.

3 St. Augustine, *City of God*, p. 109. Developments in astrophysics invite reevaluations about some so-called "archaic" *anthropocentric* scientific assumptions.

4 J.W. Yolton, *Locke and French Materialism*, pp. 1-2. Yolton refers to Professor Pierre Coste's translation in 1700 of the Locke's *Essay Concerning Human Understanding* in which Coste noted the relevant passage between himself and Locke concerning the thinking-matter suggestion. Also see Charles Taylor's *Sources of the Self* (NY: Cambridge University Press, 1989), pp. 127-42, in which Taylor discusses both Locke and the extraordinary modernity of Augustine's thought.

that Locke's thinking-matter thesis influenced French materialism. But Augustine suggested long before, in *consistency* with his own position, that "God infused into it [a body or material thing] a capacity for reasoning and intellection" as well as that a divine creative force took on "form... by the intrinsic power of the Creator who said: 'Do I not fill heaven and earth?'."[5] This thesis also involves a principle of divine intelligence in matter without imposing on reality a dogmatic norm of rationality. Accordingly, beyond a mere rationality of either Plato or Aristotle, Augustine provides a nondogmatic and distinctive basis for Locke's suggestion.

Whereas Locke's suggestion has an evident aim of making religion reasonable, Augustine holds that intelligence is infused into matter in a manner which is enigmatic to reason. His thesis of inner forms is in accord with intriguing scientific insights about phenomenal domains, as expanded on shortly, without untenably bifurcating forms from particulars. Plato's distinction between two worlds of forms and particulars led to Aristotle's criticism. To say particulars share in forms is to use poetical language when, by Plato's unqualified appeal to reason, the language begs for rational articulation.

The articulation is inapplicable to a Christian philosophy which, in not dogmatically appealing to ideals of rationality, permits a spiritual realm and mystery. Aristotle's criticism has no force against Augustine. This is not a copout in which mystery permits irrationality. Precisely, in rejecting an irrational demand that reality conform to reason,[6] he *(i)* renders viable an enigmatic willful evil of persons that was alien to Plato, and *(ii)* clarifies that what is *impossible* for intelligible forms and matter is possible for God. The two realms are breached since

---

5 St. Augustine, *City of God*, p. 256. See p. 265 for the following reference.

6 A disregard of the notion that there is nothing self-contradictory in saying that reality need not abide by the Noncontradiction Principle may be behind attempts to "make faith reasonable" and may be exacerbated by confusing a methodological rationalism, for limited purposes of science, with a philosophical one. Locke evidently imposed the latter on theology. In *Locke and French Materialism*, Professor Yolton contrasts Locke's "reasonable Christianity" to Malebranche who accepts by traditional faith the Trinity, conception, incarnation and so on.

God is wholly transcendent and immanent. This duality is clear when he notes through the Prophet:[7]

> 'I fill heaven and earth' [Jeremiah 23:24]. But this does not mean that, in heaven, we shall say that God has one part there and another part on earth... He is wholly in heaven and... on earth; and He is in both simultaneously... which is utterly *impossible*... [for] any material substance.

What is modally impossible for material substance, given its experienced nature, may be possible for God. One can say that God is simultaneously in the spiritual and phenomenal realms but that things in the latter realm cannot be in the former. And as previously noted about Augustine's thought,[8] this point is consistent with specifying contradictory notions about phenomena in the *phenomenal* realm.

Augustine's insight about the two realms may, by analogy, be illustrated with an example in modern physics. A formula in classical physics, say $P_i = m_i v_i$ for calculating the momentum of a point particle, may be applicable to both the microscopic and macroscopic worlds in an overlapping domain where Planck's constant is negligible.[9] But the formula

$$P = mv = h / \lambda$$

may only be applicable to the microphysical realm, where $P = m\ v$ = magnitude of the momentum of a moving particle, h = Planck's constant, and λ = the de Broglie wavelength that has "no analogy among all the kinds of waves studied in classical physics."[10] At the same time, this formula exhibits a duality wherein light may be interpreted as both a particle (p) and nonparticle (λ). The analogy may seem weak since the classical formula obtains only in both realms when "h" is small. But it

---

7 St. Augustine, *City of God*, p. 536. Emphasis added.

8 St. Augustine, *Confessions*, p. 318.

9 *E.g.* $P_i = m_i v_i$, where the momentum P_i of a point particle is the vector quantity equal to the product of the mass m_i of the particle and the velocity v_i of its motion.

10 *Cf.* B. Yavorsky and Yu. Seleznev, *Physics: A Refresher Course*, Tr. by G. Leib (Moscow: MIR Publishers, 1979), p. 475.

seems strong in the sense of underscoring that *(i)* a relationship between two realms, held to be conceptually distinct,[11] cannot be rejected cavalierly by a paradigm Aristotelian criticism, and *(ii)* what is deemed impossible for something in one realm or world may be possible for something in the other even in considerations of modern physics.

Is not a possibility of the "impossible", regarding two worlds, alluded to in Yolton's discussion of Locke? In the context of Locke's Christianity, divine thought is present in both transcendent and material worlds. Though the latter world is viewed classically in terms of efficient causes for an epistemic preamble to a Cause qua God,[12] a God in the two worlds begs for a Platonic ontology. The ontology accords with traditional interpretations without ignoring an Aristotelianism related to the modern philosophy of science.

The question ensues of whether, or to what degree, Augustine may have influenced Locke. It is beyond present purposes to adequately respond to the question. But it is an important one. For Locke's suggestion of a single substance with dual properties of thought and matter, besides reflecting an Augustinian teleology, may have modified French thought (the *philosophes*) more than is imagined. For example, if he did influence Locke, he may have influenced the existential phenomenology of Jean-Paul Sartre. A pre-Sartrean phenomenology in Augustine's thought goes far beyond, in sophistication, the Lockean thinking-matter thesis with a notion that matter, in thinking, is infused with consciousness.

A Modally Relevant Consciousness

In addition to either anticipating or influencing Locke's thinking-matter thesis, Augustine presaged central elements of a phenomenological epistemology. Simply stated, phenomenology denotes a branch of continental philosophy that, thriving both before and after World War II, is concerned with

---

11 Though Niels Bohr's "correspondence principle" specifies overlapping theoretical regions for a coherent single universe, a quantum or Newtonian or relativistic domain per se is still conceptually distinct.

12 *Ibid.*, p. 233.

fundamental appearances. Appearance involves consciousness and there are also investigations of it, apart from presupposed concepts, precisely to provide an ontological basis for concepts without circularly supposing them. Existentialism may be similar to an applied phenomenolgy in which implications are emphasized for everyday endeavors as well as more esoteric theoretical ones. While ignoring Augustine's Aristotelianism, S. E. Stumpf recognizes a basis for it by noting that, according to Augustine, "Any conscious person is certain... he can think, 'for we *are*... and we know we *are*, and we love our being and our knowledge of it...'."[13] Given our incontrovertible experience of ourselves, there can be no intelligible talk of either knowing or thinking apart from our consciousness of it.

For instance, a modally relevant consciousness is afforded by Augustine's insight: [14] It is impossible for "Persons do not count" $(\sim P)$ to be false when "they are not conscious of counting" $(\sim C)$ is true:

$$\Box(\sim C \longrightarrow \sim P).$$

Taken with the definition

$$\Box(\sim p \longrightarrow \sim q) =_{\mathrm{df}} (\sim p \Rightarrow \sim q)$$

and "Perons count" (P), there is a modal analogue of the sentential logic *modus tollens* for validly inferring $(/\therefore)$ "Persons are conscious of counting" (C):[15]

$$(\sim C \Rightarrow \sim P) \wedge P \;/\therefore\; C$$

13 S.E. Stumpf, "St. Augustine's Christian Philosophy," p. 137.

14 Cf. St. Augustine, *Confessions*, p. 219.

15 Since Augustine largely refers to conclusions, the symbol " $/\therefore$ " (therefore) is used for validly-drawn ones: It is impossible for the conclusion C to be false when

$$(\sim C \Rightarrow \sim P) \wedge P$$

is true because the latter logically implies the former. Cf. modal and traditional symbols in H. Kahane and P. Tidman, *Logic & Philosophy* (Belmont, CA: Wadsworth Publishing Co., 1995), p. 385.

To count is to think. In light of our immediate and incontrovertible experience of thinking, it is senseless to say we think but are not conscious of it. Modern thinkers may disparage anthropocentrism. But there is simply no way to render thought modally intelligible apart from a consciousness of it. And hence, in terms of a thinking-matter thesis, *necessarily* if divine thought is infused into matter, so is the consciousness of an Aristotelian Nous or a Judeo-Christian God who, with no pantheistic thesis, "fill[s] heaven and earth".

As our thought about matter involves a consciousness of it and its distinction from thought,[16] an awareness of thought is one of our freedom to think. Profoundly relevant to modern issues of moral and scientific truth, all things are not merely understood in terms of phenomenal or noumenal ideas. Freedom is not relegated by *pure reason*, as it is in a Kantian tradition, to a *practical idea* supposed by moral praise and blame. And even beyond this tradition, Augustine bears critically on a continental one which includes Sartre.

Sartre's strikingly similar view of consciousness is briefly addressed after briefly noting that Augustine's remarks involve not only the one in the *Confessions*. They include those in *On Free Choice of the Will* in which "To show that God exists, it is necessary to investigate man's consciousness"[17] and even "a beast is aware that... it sees when it sees";[18] the *City of God* whereby "we are certain that we possess these three things [knowledge, love, and death]... by our own consciousness of their presence in our interior and unerring vision";[19] and *On Christian Doctrine* where those who apply rules of rhetoric, when discussing things in the world, "cannot be aware of the fact that

---

16 The notion that a consciousness of reality *per se* is infused into scientific observation, undercuts a theory-dependency thesis, from Karl Popper to Paul Feyerabend, that challenges scientific realism. See R. Trundle's "Physics and Phenomenology," in *New Horizons in the Philosophy of Science*, Ed. by D. Lamb (London: Ashagate Publishing, Avebury, 1992), pp. 66-86.

17 St. Augustine, *On Free Choice of the Will*, p. 40.

18 *Ibid.*, p. 45.

19 St. Augustine, *City of God*, p. 239.

they are applying them while they are speaking unless they are discussing the rules themselves".[20]

Consciousness and Freedom

Augustine's remark about the rules is particularly insightful. To say one *cannot* be directly conscious of things about which one speaks, since one is immediately aware of one's rhetoric, is to say, given our psychic nature, that this is a modal impossibility. This brings to mind Sartre's suggested critique of a philosophical, as opposed to methodological, determinism in science. Scientists may be determinists for limited purposes of manipulating and predicting phenomena.

But scientists cannot confuse their own simultaneous modes of direct consciousness of phenomena and indirect awareness of that consciousness with the phenomena held to be causally determined: It is *as* impossible to be directly conscious of either external phenomena or internal sensations without an implicit awareness they are not consciousness *as* not to be implicitly aware of one's freedom to either be or not be a "consciousness of this or that thing-in-the-world."[21]

At the same time, Sartre's analysis is limited to human beings: We are not free to be directly conscious of things other than ourselves when conscious of such non-conscious things as our psychic states. Hence his claim that God is an impossible unity of absolute non-conscious and conscious Being, which means He cannot have absolute freedom (*libertas*), begs the question of why an impossibility for us is one for Him. In inconsistency with existentialism, and not unlike Locke, he demands that this Being conform a priori to his rational concepts. Yet, he does at least appeal to our self-conscious experience.

Augustine appeals to an experience of our mind, body and will as being both distinct and one inseparable life. He does so

---

20 St. Augustine, *On Christian Doctrine*, p. 120.

21 Cf. Jean-Paul Sartre, *Being and Nothingness*, Tr. Hazel Barnes (NY: Philosophical Library Inc., 1956), p. 316. These insights raise questions about W.T. Jones' assertion, in *A History of Philosophy*, p. 129, that Augustine had a "naive teleology that... made impossible... any real scientific advance" if the advance has been impeded by a naive determinism of modern science.

without imposing a priori rational demands on even *our* own living being. Moreover, the foregoing may reflect his influence on the notion that we are incontrovertibly aware of our free will. We are not essentially either phenomena or Cartesian thinking substances but beings aware of our freedom to think or not to think — even, paradoxically, that our thoughts are determined phenomena!

Freedom is not a matter of rational proof. Kant's relegation of freedom to a possibly real *noumenon*, since it is not a phenomenon, is as regrettable as Bertrand Russell's demand for rational arguments: "I am not prepared dogmatically to deny free will, for... there may be very excellent arguments... which either I have never thought of, or else have not had their full weight with me."[22] Especially lamentable would be Sartre's failure to develop his insight on love as a freely chosen ideal: An "ideal standard of the *value* 'love'... is in a fusion of consciousnesses in which each [lover and beloved]... preserve his otherness".[23]

Augustine's notion of an otherness of a freely loving Creator in whose image we are created, as beloved conscious beings, was highly fruitful. It led to others such as an impossibility of not being conscious of either loving or thinking and of a freedom to do them in light of a divine standard of which we are also aware. And in regard to science, a limited methodological determinism is alone warranted for such purposes as predicting phenomena. He goes beyond many elements of modern rationalism and empiricism as well as of existentialism.

It is beyond present purposes to elaborate on Augustine's contributions to existentialism. They, and his Aristotelianism, add an ignored depth to his Platonic Christianity. His insights far exceed textbook acknowledgments such as those that are only minimally noted in Kaufmann's classic commentary:

> If we look for anything remotely similar in the long past of European literature [on existentialism], we do not find it in philosophy but... in such Christian writers as Augustine... this

22 Bertrand Russell, *The Autobiography of Bertrand Russell* (Boston: Little, Brown and Co., 1967), p. 70.

23 Sartre, *Being and Nothingness*, p. 376.

> wallowing in man's depravity and... dark side of man's inner life.[24]

But Augustine's ignored insights on consciousness indicate that it is not his stress on our inner emotional life which essentially relates his thought to existentialism.

An Existentialist Essentialism

At the same time, existentialism has come to be associated with an atheistic dictum "existence precedes essence". This conflicts with our awareness of an essence of justice that is not created but may be freely sought. Appealing to our experience, Augustine speaks of our free will in relation to a higher Will. We may willfully rebel against it. But what better explains moral certainty even if societies change? There is a certainty that it is wrong to publish faulty medical research which may result in death and to commit mass murder, euphemistically called "ethnic cleansing", in modern societies. Moral traditions, rooted in a God of Nature for natural rights, have been the perennial basis for a right to life. After noting that Life is God since in Him we know no death, wherein death is experienced as unnatural, he connects an errorless knowledge to a force of love which will "meet no resistance."[25]

Knowledge of the ultimate goodness of love is from our own consciousness of its "presence in our interior and unerring vision."[26] In contrast to influential modern philosophers who hold that sentences which specify a purpose of life are senseless, this vision evokes an *a priori* purpose of loving God who loved us first, and, thus, of loving others as oneself. Love is the basis of good and does itself stem from our inner "witnesses."[27] We palpably witness ourselves as existing in the image of one God, not only by a natural trinity of mind, body, and will but of a spiritual one of love, life, and truth. Accordingly, a God of Nature is related tangibly to our spiritual nature and complements a personal God.

---

24 W. Kaufmann, *Existentialism From Dostoevsky to Sartre*, p.13.

25 St. Augustine, *City of God*, p. 239.

26 *Ibid.*, p. 239.

27 *Ibid.*, p. 239.

In sum, Augustine is sometimes eschewed by historians of philosophy for holding a puerile existentialism and by existentialists for holding a Platonic essentialism in conflict with the dictum "existence precedes essence". But he posits an incontrovertible awareness of essences, as inner forms of human nature and Nature, that may be consciously affirmed for fulfilling our existence. He articulates our self-conscious freedom beyond an Aristotelian notion of our being causally inspired to strive for natural ends (*entelechy*). And he explicates the ends in terms of an existentialist essentialism that stresses an awareness of free will as a condition for "responsible living."[28]

A modernism of his existentialist essentialism is clear by a comparison to Kierkegaard's essential and existential human conditions. Interestingly, Malcolm notes that Kierkegaard and Augustine were the thinkers most admired by Wittgenstein. And Hanna Pitkin holds that "Wittgenstein cherished what the existentialists came to call 'authenticity'— the willingness... to live by what you say."[29]

This living is possible, in Augustine's thought, by being aware of divine truths on the basis of experiencing our inner nature and outer visible Nature. And the living bears on the practical relevance of morality. In view of a moral crisis in modern scholarship, for example, Susan Haack argues for an integrity that is "epistemologically and morally valuable."[30] Yet the value of integrity is undercut in an anti-metaphysical tradition by a deontological and consequentialist ethics. One involves our rational nature and the other consequences that are good. Thus, there is begged for a goodness of our nature and things in Nature. But their Kantian relegation to mere possible realities was aggravated by a verification principle. Truth-claims about the realities, and hence what is valuable, would amount to either metaphorical claims or metaphysical sentences without meaning.

---

28 *Ibid.*, p. 110.

29 Norman Malcolm, *Ludwig Wittgenstein: A Memoir* (NY: Oxford University Press, 1984), p. 59, and Hanna Pitkin *Wittgenstein and Justice* (Berkeley: University of California Press, 1974), p. xi.

30 Susan Haack, "Concern for Truth: What it Means, Why it Matters," *Annals of NY Academy of Sciences* 775 (1996) 57-63, abstract.

Chapter 5

Theological Modalities and Meaningfulness

The logical positivism of A. J. Ayer's *Language, Truth, and Logic* construes assertions to be meaningless when there is no place in the empirical world "to which they refer."[1] To affirm Ayer's philosophical contributions is not to deny his pejorative influence on an overly restrictive view of human experience. It is a foregone conclusion that any sentential references to non-empirical good and evil things, much more to God and life after death, would be meaningless.

Ayer's verificationism, as a criterion for meaningfulness, had an interesting outcome. He came to acknowledge that there is life after death, even if it is not empirically verifiable, in his startling 1989 article "What I Saw When I Was Dead". Though predictably dismissed *a priori* by many of his Anglo-American followers who permit nothing to count against their own interpretation of "experience", the latter is examined in terms of Augustine's broader understanding as a springboard for amplifying his theological modalities.

Verification and Meaningfulness

The positivist notion of meaninglessness has an uneasy relationship to science. The uneasiness stems from a verification principle that, though initially formulated for science, came to

---

[1] A. J. Ayer, *Language, Truth, and Logic* (NY: Dover Publications, 1952), p. 43. Certain sentences are viewed as paradigm instances of meaning-lessness. These sentences include ones which make reference to spiritual realities and to Heidegger's notion of "nothingness".

be applied dogmatically to *all* sentences. Without any qualification, it specified that sentences are meaningful if and only if they are empirically or logically true (false).[2]

Now this conservative verification principle was superseded by a liberal one that involves empirical evidence merely "counting for or against" sentences or theories. Despite its being more "tolerant" even if no evidence counted for or against *it*, Hilary Putnam notes that ordinary persons as well as philosophers have tended to employ, indiscriminately, the conservative version in virtually every human endeavor:[3]

> The forms of 'verification' allowed by the logical positivists... have been *institutionalized* by modern society. What can be verified in the positivist sense... exemplifies and reinforces images of knowledge and norms of reasonableness maintained by culture.

The influential positivists of Anglo-American culture suggested that sentences of the form "*P* is good" or "*R* is evil" amounted to "I like *P*" or "I do not like *R*". The sentences were either meaningless or, at best, relevant to a science such as behavioral psychology. There can be little doubt that a transposition of moral claims into psychological ones, or ones that designate mere personal taste, has been largely institutionalized in many Western societies.

For example, Professor A. Baier, in "Parents, Schools, and Values Clarification," notes that "since 'by definition and right... values are personal things' [per the seminal teacher's guide *Values and Teaching* (1966)], teachers should never try to teach children correct values."[4] Baier further notes that the influential educational authors Sidney Simon and Howard Kirschenbaum have convinced many administrators that teachers should not "'moralize, preach'... and rely on 'religion

---

2 The principle is evidently meaningless when applied unqualifiedly since it is not itself either empirically or logically true (false).

3 Hilary Putnam, "Philosophers and Human Understanding," *Scientific Explanation*, Ed. by A.F. Heath (Oxford: Oxford University Press, 1981), p. 100.

4 A. Baier, "Parents, Schools, and Values Clarification," *The Wall Street Journal*, April 12, 1982.

and other cultural truisms'."[5] Here, there are the seeds for a pejorative sort of "multiculturalism". He adds that Professors Kenneth Strike at Cornell University, Alan Lockwood at the University of Wisconsin, and John Steward of Michigan State are among the few who have had the courage to challenge values clarification. In addition to its potential for politicizing "truth", they warned of its being a euphemism for narcisstic self-gratification and for indoctrinating students with an ethical relativism.

That is, if students were being *educated* to critically think, they would presumably consider something as patently unethical as teachers who discriminate against students on the basis of their "looks". If a value for the discrimination was formulated as a moral claim that may be true for some teachers and false for others, where the truth-condition for "truth" is the opinion of any given teacher, the students might see that values clarification permits the moral claim to be incoherently both true and false. Also, given that truth-value per se is relative, the clarification theorists would assume at a meta-ethical level that ethical theories are also relative.

Students who enjoy an education, which fosters critical questions, might question if this metaethical assumption is itself both true and false. It might be true for values-clarification advocates and false for others. Surely, the assumption is false for others. Thus, students may wonder why, in view of associated agenda such as "cultural diversity" and "life-style tolerance", the advocates intolerantly insist on their own metatheory. An indication that both metatheory and theory are relativistic is indicated by a general reluctance to refer to "truth" as opposed, say, to "correct". Understandable misgivings about "political correctness" have themselves been politicized as concerns of either *unenlightened* "right-wing" students or their parents.

Ethical relativism is not new. It has a mischievous origin in Greek Sophism, in the fifth century B.C., which is identified with an ancient "Enlightenment" by adherents who take seriously the historical designation. The adherents tend to be

5 *Ibid.*

hostile to an "unenlightened" traditional religion and view relativism as properly undermining it. Ironically, though they paved the way for the belief that religion should be superseded by science, scientific objectivity came under attack as well. In leaning on such examples as water feeling cold to some persons and hot to others, the Sophists held that a given scientific claim, by analogy, could be both true and false.

The Sophists tried to obfuscate the fact that internal sensations do not counter objectively measurable phenomena or states of systems in terms of such things as temperatures. Augustine noted that "superficial questioners and careless students..., these loquacious sophists, are also in the habit of harassing the faith of the less learned".[6] While philosophers have both flirted with and denounced sophistic relativism, he also refers to the Scripture in which it is stated "He that speaketh sophistically is hateful."[7]

Science Paradigm for Experience

Most modern philosophers are not relativists and may *believe* in objective moral norms. Nevertheless, many notable philosophers and scientists have subtly imposed a verificationist norm of thinking on moral truth-claims. This further illustrates how the verification principle has been institutionalized as a norm of reasonableness. A reasonableness of both applied ethics and ethical theory has tended to be weighed by their compliance with "neutral territories" of reason and sense experience. And the latter are expressed ideally by scientific descriptions.

This ideal amounts to a virtual setup for a Naturalistic Fallacy wherein prescriptions cannot be deduced from scientific descriptions. Appeal to the Fallacy is predictable when one archetypal ethics text states that ethics "is like the experimental sciences... studying relationships of empirical data".[8] And ties to the data are made more worrisome by appeal to

---

[6] St. Augustine, *On Free Choice of the Will*, p. 111.

[7] St. Augustine, *On Christian Doctrine*, p. 67. See Ecclesiastics 37:23.

[8] M. A. Gonsalves, *Fagothy's Right and Reason: Ethics in Theory and Practice*, 9th Ed. (London: Merrill Publishing Co., 1989), p. 193.

scientific "knowledge by which moral convictions can be shown to be true or false."[9] The convictions may be compared to scientific knowledge in order to mimic objectivity. But modern philosophy has abandoned any theological metaphysics of a Creator, and of a divine element in human nature and Nature, that makes them good. Thus scientific descriptions of our nature are devoid of any prescriptive import.

It is beyond present purposes to survey a virtual plethora of literature that disjoins descriptions of our nature from how it ought to be fulfilled. One need only consider a sample of some notable scholars who, while having made admirable contributions to modern thought, suggest the separation.

Richard T. DeGeorge, a former President of the American Philosophical Association, denies any role for traditional metaphysics, much less for Augustine's theological ethics. For ethical issues should be addressed by neutral territories of "reason and human experience alone... [without consideration of] God's existence or of revelation".[10] Social scientist Steven Goldberg is more direct. He says that "Science, the explanation of what *is*, can never... generate a view of what *ought* to be."[11] And whereas Abraham Edel laments that moral philosophy is "not much further advanced than the Pre-Socratics were in physics",[12] philosopher E. M. Adams makes religious discourse "accountable to science" by a "tolerant" verification principle in which that discourse is either metaphorical or false.[13]

Finally, physicist M. A. Rothman decries that a Judeo-Christian explanation of God as Creator is "not even an empirical statement."[14] And philosopher Thomas O'Dea notes,

---

9 *Ibid.*, p. 193.

10 See R. DeGeorge's anti-theological "Theological Ethics and Business Ethics," *Journal of Business Ethics* 5 (1986) 424 as well as R. Trundle's response "Is There Any Ethics In Business Ethics?," *Journal of Business Ethics* 8 (1989) 261-69.

11 S. Goldberg, "Letters," *National Review* 47 (1996) 2.

12 A. Edel, "Romanel Lecture," *Proceedings of the American Philosophical Association* 60 (1987) 823-40.

13 E. M. Adams, "The Accountability of Religious Discourse," *International Journal of Religion* 18 (1985) 3-17.

14 M. Rothman, *A Physicist's Guide to Skepticism* (NY: Prometheus Books, 1988), p. 161.

with no reservations, that science has tended to "mathematize thought and has consigned [religious and moral] conceptions which cannot be thus refracted to a limbo of second-class importance."[15]

How do the foregoing remarks, influenced by a verification principle, beg for a reexamination of Augustine's insights? His notion of faith is significant. Without becoming narcisstic or cynically relativistic, faith may strengthen the resolve of philosophers and scientists to acknowledge that moral "reasons" can only be logically true by being trivial and that moral prescriptions cannot be deduced from descriptions of modern science *per se* — especially when there is dogmatic appeal to a verification principle.

In addition to the principle not being *itself* verifiable, the tenability of faith is tied to its consistency with both empirical science having such purposes as predictions and there being other meaningful purposes in life. Augustine, in speaking of astronomy, notes that it yields "predictions concerning the future" but is only relevant to gaining "information that pertains to the stars themselves."[16]

Augustine's misgiving that considering only "Knowledge of this kind... is associated with the most pernicious error"[17] seems well warranted in view of willfully erroneous uses of the principle. Putnam suggests that many philosophers have continued to use the principle indiscriminately even after knowing it was self-refuting. "Strangely enough," he notes, a knowledge that the principle was self-refuting "had very little impact... I want to suggest that the neglect of this particular philosophical gambit was a great mistake."[18]

A mistake? A "gambit" is a clever maneuver or ploy. The ploy leads to the view that an *intellectual dishonesty* of willfully misusing the principle would amount to nothing more than the subjective peculiarity of disliking something such as

---

15 Thomas O'Dea et al, "Religion in the Year 2000," *Philosophy Looks to the Future* (Ill: Waveland Press, 1985), p. 545.

16 St. Augustine, *On Christian Doctrine,* p. 66.

17 *Ibid.,* p. 66.

18 Putnam, "Philosophers and Human Understanding," p. 100.

dishonesty. An assertion that the *dishonesty* was immoral would amount to a *dislike* of the self-refuting criteria for meaningful truth-claims! How are the claims connected to understanding an ordinary experience of love, stressed by Augustine, in the human sciences?

Human Sciences and Love

Taking physics as their model with our caveat that their observation terms have not been incorporated into theories with predictions comparable to physics, the human sciences have either excluded "love" as a definiendum, since there is no clear definiens, or included it with a reductive definiens. That is, "love" is typically defined as a sex drive that is represented classically as the "id". In distinguishing ambiguous terms responsive to clarification from vagueness, one logic textbook states that whereas words such as "love" are vague since we cannot "tell with any... precision whether they apply to a given situation," scientific terms such as "'acceleration' in Newton's second law [are] more or less fruitful, depending on the deductive consequences".[19]

But does not unconditional love of husband for wife and they for children, for instance, have common-sense as well as empirically evident consequences? Having been excluded from scientific terminology because it is not an observation term that can be transposed into a theoretical term, say as "water" into "H_2O", agape love as an undeniably experienced reality is neglected by the very human sciences which may exacerbate the pathologies their therapies portend to ameliorate.

"It doesn't take a Ph.D. in sociology or psychology," says social scientist W. A. Donohue, "to know that... those who should know better, namely, students of human behavior, often provide advice that makes the situation worse."[20] In noting that a human-science culture of the "id" contributed to a sexual revolution in the 1970s, he speaks of a "scientific" approach

---

19 P. J. Hurley, *A Concise Introduction to Logic* (Belmont, CA: Wadsworth Publishing Co., 1991), pp. 86, 88.

20 W. A. Donohue, *The New Freedom: Individual and Collectivism in the Social Lives of America*, p. 67.

that ignores an accompanying divorce epidemic and epidemics of virulent sexual diseases, children born out of wedlock, sky-rocketing juvenile crime, and a feminization of poverty:[21]

> Here's how it works. We are introduced to a long list of primitive cultures... where sex is free and open... Of course, many of these liberated peoples also practice cannibalism, subject women to clitorectomies, and decapitate their adversaries, but this is not to count... [and] when the pitch for another round of tolerance is made, the discussion typically focuses on the nefarious influence that religion has exerted... To anyone so naive as to believe this diatribe (the intended audience is most often college students), the logical inference... is that religion enslaves... the rejection of it enlightens.

Clinical psychologist L. Crabb as well as psychiatrists F. Minirth, M. Nicoli and W. Wilson, are among a growing number of researchers who underscore that enlightened science does not reject religion. This is not a matter of "Bible pounding" because the vagueness of "love" must be weighed against its effectiveness in coordinating behavioral diagnoses with prognoses or therapies that ameliorate psycho-physical pathologies.

The verifiable success of diminished pathologies *is* the scientific methodology for corroborating theories that portend to reflect human nature. The success of theology-oriented therapies does bear on our nature and has evidently been known, for some time, to compare quite favorably to secular results. In addition to the work of Nicoli and Wilson in *The American Journal of Psychiatry* and *Diseases of the Nervous System*,[22] Crabb notes a stream of consciousness in his patients that bears poignantly on the issue:[23]

> I must be loved by another person, loved unconditionally, accepted just as I am without demand or pressure. If I am loved

---

21 *Ibid.*, p. 98. Augustine is not naive about sex: "Men declare that they are happy when they embrace beautiful bodies..., whether of their wives or of prostitutes; do we doubt that we are happy in the embrace of truth?" See *On Free Choice of the Will*, p. 67.

22 See A. M. Nicoli, "A New Dimension of the Youth Culture," *American Journal of Psychiatry* 131 (1974) 396-401; W. P. Wilson, "Mental Health Benefits of Religious Salvation," *Diseases of the Nervous System* 33 (1972) 382-386.

23 See among L. Crabb's other works, *Effective Biblical Counseling* (Mich: Zondervan, 1977) 61.

because I behave well, I am under pressure to keep on behaving well. I know I might not. Therefore I could lose love. I must be loved with an acceptance... I cannot lose no matter what I do.

Though what "I do" does not exclude God's judgment, it does not change His love. Love (*agape*) is not only tied to our incontrovertible experience but to a God who, being an immanent and transcendent principle of Life, bears on issues of death. Consider various modalities in terms of experience.

Experience and Deontic Modalities

A. J. Ayer's article "What I Saw When I Was Dead" should be disconcerting to many of his Anglo-American followers.[24] For in addition to his well known commitment to a Humean empiricism which disallowed immaterial realities and reduced thinking to a physical species of sensation, he was one of the most outspoken twentieth-century critics of both religion and life after death.

Augustine refers to similar philosophers whose "criterion of truth is the bodily senses, and who would have us believe that all knowledge is to be measured... by such deceitful testimony."[25] He adds that when they speak of something beautiful in their unguarded moments, "I often wonder by... what bodily senses they have perceived that beauty".[26] The point is that science-oriented empiricist philosophers tend to

---

24 A. J. Ayer, "What I Saw When I Was Dead," reprinted by arrangement with the *London Sunday Telegraph* in the *National Review* 40 (1988) 38-40. Sentential affirmations of "life after death" as positivist paradigms of meaninglessness have not prevented scholarly analyses of reported "Near-Death-Experiences" (NDE). See R. Moody, M.D., Ph.D., *Life After Death: With a Foreward by Elizabeth Kubler-Ross,* M.D. (NY: Bantam Books, 1995). The Kentucky Board of Nursing, among professional conferences, organized an NDE Conference on 20 February 1988 with thirteen NDE speakers. One of them, Professor Howard Storm, became a minister after his self-professed atheism. Some philosophers object that the experiences beget inconsistent religious interpretations in different cultural contexts. But a possible initial "cultural norming," in an after-death spiritual realm, would be no more astonishing than the reality if skepticism is momentarily suspended. Philosophically, which reality obtains would still be an open question.

25 St. Augustine, *City of God*, p. 154.

26 *Ibid.*, pp. 154-55.

employ the ideas they reject, if not in their philosophy, then in their ordinary lives. Ironically, Ayer's allowance for *unverifiable* ideas became public by his own efforts after his self-avowed death.

After Ayer said he had died in a London hospital,[27] he held that he was incontrovertibly conscious. Later, he noted the "strong evidence that death does not put an end to consciousness."[28] Aside from the senselessness of reducing a consciousness *of* one's sensations to sensations, there are ordinary distinctions between sensations of pleasure derived from tasting something good and a good deed as well as between the pain of a stubbed toe and guilty conscience.

If positivists permit such everyday common-sense distinctions by abandoning a self-refuting verification principle, they break with the empiricist tradition in a way that makes it difficult to see how they could reject a priori Ayer's claimed experience. Said by medical personnel to be lifeless, his body was on a bed when he says he experienced rational thoughts and was "aware that this [painful *red*] light was responsible for the government of the universe."[29]

Ayer's recognition of a universe, unverifiable by sensations in this world, brings to mind Augustine's assertion that the right use of death, originally "intended as the punishment of sin, may bring forth abundant fruits of holiness."[30] While dying is not good in itself, experience of it in terms of ourselves or others underscores its being "harsh and unnatural".[31] Hence, the experience draws attention to something natural about eternal life that is a short step from a more liberal

---

27 To an objection of there being various criteria for death whereby there is a questionable verification of Ayer's, one obvious response is that Ayer was a world-renowned philosopher who pioneered criteria for verification. His own unequivocal use of the word "death" and articulate descriptions, made in careful reflection, might render presumptuous the objection.

28 Ayer, "What I Saw When I Was Dead," p. 39.

29 *Ibid.*, p. 39. Whereas Ayer stressed the experience of a painful red light, other persons felt no pain in the presence of a white one.

30 St. Augustine, *City of God*, p. 276.

31 *Ibid.*, p. 275.

construal of experience to a "self" that survives bodily death. The survival finds expression from Socrates' *Apology* (40d-41c) to Wittgenstein's remark that whatever happened to him "externally", it was *impossible* that anything harm his "innermost being".[32]

"Life can educate one to a belief in God," said Wittgenstein who expressed awe of Augustine and added that "experiences too are what bring this about... sufferings of various sorts."[33] The suffering may itself be a short step to belief in a resurrected body since the unnaturalness of death, notes Augustine, is precisely "the violent sundering of what, in a living person, were so closely linked and interwoven;... soul and body."[34]

In short, Augustine has no illusions about a perennial deceitfulness of certain philosophers: If Ayer's followers were unmoved by their own self-refuting verificationism, they would be untouched by his experience. Still, those who are not blinded by pride, in Augustine's terminology, would not reject a priori the accounts of life after death. A serious consideration of the accounts does not mean that he would glibly accept "near death experience" but rather, arguably, that he would allow for its possibility.

Faith in life after death is rooted in earthly experience. As an experience of our biophysical nature leads to scientific necessities or impossibilities that are not generally dismissed by mere empirical facts, an equally viable experience of our psycho-spiritual nature may permit religious and moral claims about what either *must* or *cannot* be the case which avoids a Naturalistic Fallacy.

Religio-moral claims are not inferred from mere physical descriptions. Arguably, they may stem from inner natural norms of human nature, that are themselves derived from an intrinsic power of God who orders all things,[35] and are complemented by revealed norms of Nature's transcendent God. Regardless of

---

32 Cf. new access to Wittgenstein's diaries in R. Monk's *Ludwig Wittgenstein: The Duty of Genius* (London: Penguin Books Ltd, 1990), p. 116.

33 *Ibid.*, p. 572.

34 St. Augustine, *City of God*, p. 275.

35 *Ibid.*, p. 265.

disbelievers, allegations of a Naturalistic Fallacy seem moot: From the sacrament of marriage in which a wife says "My husband *cannot* leave me for a younger woman," and an irrelevant objection "But he did!",[36] to "Unless a man be born again...he *cannot* enter into the kingdom of God"[37] and the extraneous objection that God's kingdom cannot be deduced from empirical descriptions.

Senseless appeals to such descriptions, as discussed shortly, bear on Wittgenstein's talk about his own sense of guilt. There are experiences of guilt and redemption addressed uniquely in Scripture. Augustine's foregoing reference to Scripture yields the impossibility that "One does not enter the kingdom of God" ($\sim G$) is false when it is true that "One is not born again" ($\sim B$). Recast as a necessity, there is the conditional:

$$\sim B \Rightarrow \sim G$$

Despite Augustine's caveat about unbaptized persons who confess Christ, which might be another form of being born again, the modality may illuminate a traditional mode of theological reasoning when taken with "One is not born again" ($\sim B$):

$$\sim B \wedge (\sim B \Rightarrow \sim G) \; / \therefore \; \sim G$$

While a conclusion that necessarily one does not enter the kingdom of God ('N$\sim G$') would commit a modal fallacy, the truth of $\sim B$ might be accepted on the basis of a self-avowed acceptance of not being born again — if not on overt behavior — and $\sim G$ cannot be false when $\sim B$ and $\sim B \Rightarrow \sim G$ are true. In sum, the modal status of $\sim B \Rightarrow \sim G$ excludes its falsity in the sense in which may be false, say, speculative theological claims. The other premise $\sim B$ and conclusion $\sim G$ admit of the logical possibility of being false and are not modally true. But this admission does not obviate a believer's reasonable belief that they are true.

---

36 *Ibid.*, *cf.* p. 319.

37 *Ibid.*, p., 275 (John 3:5, emphasis added).

Thus the Church does not traditionally decree that *necessarily* given ordinary individuals will not go to heaven. The decree would reflect a modal fallacy in the above formulation. Nonetheless, the Church may reasonably hold that some individuals will not and that, certain revelations being modally true, there are epistemic degrees of conviction.

One of the strongest convictions concerns life after death. Presumably, the conviction may have as much certitude, in virtue of our palpable sensibilities, as the virtual certainty that a man cannot morally leave his wife for a younger woman when he made a marital vow of fidelity.

Wittgenstein's "Awe" of Augustine

Augustine's Platonism discerns a spiritual world from a fallen one in which pride is the cardinal sin. It is significant in this sense, as noted by Norman Malcolm, that Wittgenstein accepted a "conception of God insofar as it involved one's *awareness* of one's own sin and guilt".[38] For he was an esteemed pioneer of analytic philosophy into which the logical positivism, he equally influenced, largely evolved.

Many of Wittgenstein's followers, whose disputes over his posthumous papers were as petty as he "was high-minded and profound,"[39] relegated religion and traditional philosophical theology to confused seminarians. However, Wittgenstein said he began his *Investigations* by quoting Augustine's *Confessions* "not because he could not find the conception expressed... by other philosophers, but because the conception must be important if so great a mind held it."[40]

Some of the things Augustine held, to which Wittgenstein alluded, are that it makes sense to say "I will, but my body does not obey me" but senseless to assert "my will does not obey me."[41] And one of Augustine's ignored contributions to analyzing

---

38 Malcolm, *Ludwig Wittgenstein: A Memoir*, p. 59.

39 See E. Toynton's interview of philosophers in "The Wittgenstein Controversy," *The Atlantic Monthly* 279 (June 1997) 28-41.

40 Malcolm, *Ludwig Wittgenstein: A Memoir*, p. 59.

41 Ludwig Wittgenstein, *Philosophical Investigations*, Tr. by G. E. M. Anscombe (NY: The Macmillan Co., 1958), p. 161.

the language of ethics, reflected in Wittgenstein's thought, is precisely a linkage of ethics to an obedient will; a will that cannot be understood to obey me but rather the commands of a source other than myself to whom I will to be faithful. Thus, says Augustine, one "is capable of comprehending a command, *obedience* to which is the first duty of faith, so that he can do what he is commanded to do."[42]

There may be, in this sense, an ignored epistemological depth of Scripture. The depth includes a recognition of morally relevant feelings of conscience ignored by secular scholars. Swayed by rationalistic elements of a Platonic-Aristotelian tradition, these scholars might lead us to suppose that we are to feel irrational rather than guilt about misdeeds. And beyond this, the depth includes an ethical language of commands. In either Aristotelian or standard propositional logic, "truth" is not ascribed to commands but to statements.

Accordingly, the Scripture, to which Augustine appeals, excludes pretensions to "moral reasons" that owe their status as reasons to either our rational nature (deontology) or desirable consequences (consequentialism). A modern restriction to empirical and logically necessary truth raises dilemmas. Insofar as the reasons are logically certain, they yield no information about our behavior and insofar as they do, they have no logical certitude. *Pari passu* reasonable persons can disagree and the reasons become relative to possibly inconsistent views.

Wittgenstein illustrated both impatience with such views and an ethical dimension of commands by relating a story. He had taken lodgings in a preacher's home and was obviously weary. When the lady of the house asked impassively if he would like this or that, the preacher commanded "Do not ask, *give!*"[43] After noting that this made a deep and morally favorable impression on him, his biographer ties him to an Augustinian sort of pessimism about the human condition:[44]

> A characteristic remark that Wittgenstein would make when referring to someone who was notably generous or kind or

42 St. Augustine, *On Free Choice of the Will*, p. 144.

43 Malcolm, *Ludwig Wittgenstein: A Memoir*, p. 52.

44 *Ibid.*, p. 52.

honest was 'He is a *human being!*' — thus implying that most people fail even to be human.

Wittgenstein's awe of Augustine's "great mind" was matched by his feeling "that our lives are ugly and our minds are dark — a feeling that was often close to despair".[45] He did not ameliorate the despair through Christianity since he regarded many of its beliefs as "based on qualities of character and will that he himself did not possess."[46] However, this fact did not erase his deep respect for the belief.

When he asserted in *On Certainty* that there is not always a clear boundary between the factually unreasonable and the logically impossible,[47] he more than hinted at an impossibility of what is modally *more* than unreasonable to believe. His distinction did not cast into question religious belief but rather appeals of disbelievers to scientific objectivity:[48]

> what men consider reasonable or unreasonable alters... is there no objective character here? *Very* intelligent and well-educated people believe in... the Bible, while others hold it is proven false, and the grounds of the latter are well known to the former.

In the end, Wittgenstein appreciated how Augustine's words bore the stamp of an extraordinary human being and how the Bible gave singular expression to his own sense of guilt. Augustine contrasted the Bible's depth to "learned persons" about whom Wittgenstein also had grave misgivings.

After reading Sir James Frazer's *The Golden Bough* which likened a "savage people's" magic and religious rituals to primitive science, Wittgenstein snapped that Frazer's own scientific outlook was incomparably more shallow: Those people will never "be so far from any understanding of spiritual matters as an Englishman of the twentieth century."[49] In preserving what was deep about the rituals, a *morphology* of

---

45 *Ibid.*, p. 60.

46 *Ibid.*, p. 60.

47 Ludwig Wittgenstein, *On Certainty*, Tr. by G.E.M. Anscombe & G.H. von Wright (NY: Harper & Row Publishers, 1972), p. 60, # 454.

48 *Ibid.*, p. 43, # 336.

49 Quote of Wittegenstein in Monk, *Ludwig Wittgenstein: The Duty of Genius*, p. 310.

them would itself have a quality of magic. By the same token, in preserving what was deep in the old metaphysical theories, Wittgenstein hoped "his new method of philosophy... would itself have the character of metaphysics, without attempting the conjuring tricks of the *Tractatus*." [50]

Though Augustine was indifferent to proofs of God, as was Wittgenstein, Thomas fulfilled a metaphysical tradition of philosophical theology by establishing the importance of such proofs. He did not merely relate the proofs to perennial scientific principles. He tied the principles to ethics, and the ethics to politics as well as answered a question that haunted Wittgenstein: Why is there a world rather than nothing? Is the question a senseless metaphysical one without a profound basis in ordinary experience? An answer begs for Thomas' idea, implicit in Greek philosophy, that the profane and sacred realms are entangled.

---

50 *Ibid.*, p. 311.

Chapter 6

Thomas' Realms: Sacred in the Profane

St. Thomas' theology involves a strong relation of the sacred and profane realms in terms, among other things, of its emphasis on a mutuality of God's will and reason being manifest in Nature. This does not conflict with a distinction of heaven from earth. It underscores that we may carry on our lives without thinking about God and, even without faith, infer His existence from an experienced dependence of all things. In virtue of their being causally dependent, individually and collectively, we might infer that nothing could exist unless there exists a Creator who is not dependent.

A modern logician may object that it is possible for a causally dependent sequence to proceed ad infinitum into the past. But the objection may not merely confuse a *logically* possible regress with a *modally* impossible one based on experience.[1] The objection may reflect a logician's disregard of *(i)* experiential truth, except for hypothetical issues of soundness, and *(ii)* a restrictive assumption biased by a Humean-positivist influence regarding the nature of "experience".

Thomas may have articulated modes of an experiential reasoning that illuminates why some eminent modern philosophers, who did not believe in a "Maker", have expressed astonishment in the form of a question. If there is a haunting

---

1 See R. Trundle's "Thomas' 2nd Way: A Defense by Modal Scientific Reasoning," *Logique et Analyse* 146 (1994) 145-168. In this paper it is argued that "modal scientific laws," based on an experienced nature of phenomena, strictly imply the Thomistic notion that it is modally impossible there are phenomena when there is no First Cause.

question of why the world exists, there may be implicit recognition that its existence is impossible when there is no Maker or Creator. Understood as a First Cause, the Creator would be a God and the philosophers are disbelievers. Accordingly, the disbelief results in astonishment over the existence of anything. Why is there not "nothing"?

Certainly, the Cause is not identical to a Judeo-Christian God. But it may be too close for comfort. The discomfort may lead to a rejection of the Cause and, therefore, to both astonishment and a specter of *nothingness*. Before examining the modal reasoning in greater depth, which bears on science as well as on ethics and politics, there is attention to Thomas' general philosophical theology. How might theology and philosophy overlap in terms of the recognition of a divine Being? Though this Being is viewed differently in Thomas' five proofs of God, an interface which is initially considered addresses only a divine Cause for the sake of simplicity.

Theology and Philosophy

Whereas Augustine viewed philosophy and theology as essentially the same thing, Thomas stressed a distinction of faith from reason. Surely, reasoning about the world, evident in many ancient Greeks, seemed to identify theology and philosophy. Aristotle's first book of the *Metaphysics* stated that the history of philosophy, beginning with Thales, was a search for both the principle cause of things and a divine Wisdom or God (a *Nous* whose thinking is a thinking of thinking). Thomas would not deny that divine thinking is tied to theology when it is understood as *natural* theology.[2] The theology is located where Judeo-Christian theology and philosophy overlap:

---

[2] St. Thomas, *Summa Theologica,* I, 1, 2. Unless otherwise noted, references to this work are from *A Summa of the Summa*, Ed. by P. Kreeft, Tr. by The Fathers of the English Dominican Province (San Francisco: Ignatius Press, 1990).

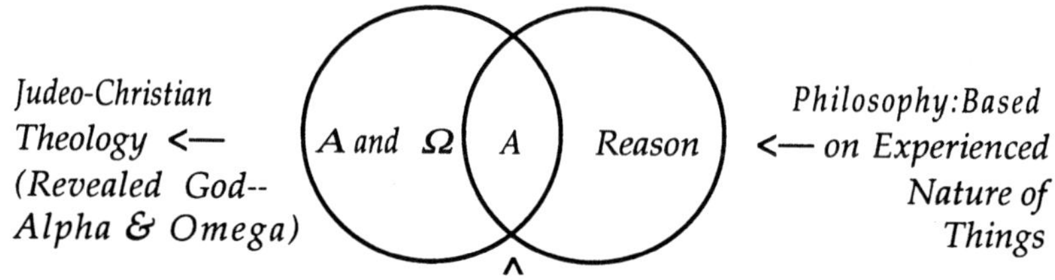

Natural Theology (Center):
Nature's First-Cause Creator (A) as God

Overlapping Theology & Philosophy

Philosophy may not be the same thing as Judeo-Christian theology. But it is a short step to the theology when one begins with an experienced Nature and reasons to Nature's God. While no faith is needed, this God evokes awe by virtue of being a creator who alone is not dependent .

The above circles distinguish philosophy from theology. This contrasts to their identification by Augustine and may suggest that he is optimistic and Thomas pessimistic about the human condition. Afterall, the human condition might not be improved by philosophizing when the latter does not involve theology. But what actually bears on their outlooks, in this context, are different emphases. Augustine tends to emphasize that since the time of the Fall, the human condition reflects a fallen nature of our mind, body, and will in which our freedom (*liberum*) of will is naturally rebellious. Our will is separated from a divine freedom (*libertas*) and, without grace, skews our intellect. Thomas stresses that intellect and will are complementary in virtue of their divine mutuality.

He says (I, 16, 5) that the "will and the intellect mutually include one another: for the intellect understands the will, and the will wills the intellect to understand." Our understanding by its nature, though not necessarily by nurture, is open to faith in the revealed God by way of a God of Nature and human nature. This view of our nature explains a Thomistic Catholicism's stress on such things as education and scientific research, the coordination of research with theological encyclicals, and tolerance in principle for many cultural perspectives — from Africa to India to Europe — assimilated into one church mass.

The *many* cultures share *one* experienced world. In being the foundation for reasoning to God, one world became the basis for five proofs of God's existence.

Essence of Five Proofs

When Thomas said nothing comes from nothing (*ex nihilo nihil fit*) and that all things have causes, which bear on several proofs of God's existence, he did not suppose a restrictive notion of modern causality. This notion involves a purposeless sequence of exactly or inexactly determined events. That is, for example, interpretations of a modern causal principle may involve formula of either classical mechanics that are deterministic of exact measurements or quantum mechanics deterministic of inexact probabilistic ones. Whereas many modern philosophers of science reject both interpretations, because their *a priori* status means that nothing counts against their truth and their truth does not follow from experience *a posteriori,* Thomas evidently embraced a modal modal truth considered later.

In order to avoid a straw-man misrepresentation of Thomas' proofs, there needs to be a proviso that he liberalizes some of Aristotle's four causes. Among them, only narrowly interpreted efficient causes were given "critical expression" by Hume and Kant during the scientific revolution. Thomas' notions include: *(i)* the *efficient causality* of a Creator as a voluntary external cause of created beings that, while not themselves creating ex nihilo, produce motion and new things out of existing ones *(ii)* the *final causality* of a Creator who is an external cause of created beings that have natural ends coordinated with the supernatural one; *(iii)* the *formal causality* of created causes intrinsic to matter that, *qua* forms, reflect essential natures; and *(iv)* the *material causality* of created causes that are intrinsic to the world as the matter out of which things are made. Having given these working definitions, the following summarize his proofs:[3]

---

[3] Cf. the "ways" in the *Summa Theologica.* Some modal terms in the first three, bearing on modal interpretations, are italicized. I am indebted to Professor Emeritus Norris Clarke, S.J., for clarification of these causes.

1. Motion: Our senses show that some things are in motion and that it is caused by other things so that, if things are at rest, they will not move unless moved by other things. Since moving things are actually in motion and things at rest only potentially, motion is the transformation of potentiality into actuality. And since things cannot attain actuality except by things already actual and it is impossible for them to be simultaneously actual and potential in the same respect, it is equally impossible in this respect for them to be both movers and moved in the sense of moving themselves. They are moved by other moved movers and they by others whereby it is *impossible* to go on to infinity since, in there being no unmoved first mover, there would be no other mover; as, all things equal, it is impossible for a staff in itself to move in one's hand when the hand does not put it into motion. Therefore, it is *necessary* to arrive at an unmoved first mover who everyone understands as God.

2. Efficient Cause: We experience an order of efficient causes whereby there is no case known, nor is it possible, for a thing to be the cause of itself since it would have to be prior to itself. But whether an intermediate cause be one or several, it is *impossible* for the order to go on to infinity since the first cause is that of intermediate and ultimate second causes. That is, if it is possible for there to be an infinite regress, there would not only be no first cause but no ultimate effect or intermediate second causes. Now to take away the cause is to take away the effect. And since it is *impossible* for there to be an effect when there is no cause, if there is no first cause among efficient causes, there will be no second causes. There are second causes. Therefore, there is a first cause which everyone names God.

3. Possibility and Necessity: Since we experience things as coming into and going out of existence, it is possible for them to be or not be. It is *impossible* that all things should always be such (or exist in such a way) since that for which it is possible to *not be* does not have being at some time.[4] So, if it is possible for everything to *not be,* then at some time there would be a time when nothing exists. If this possibility obtained, there

[4] *Cf.* P. Kreeft, *A Shorter Summa* (San Francisco: Ignatius Press, 1993), p. 61: "I.e., given infinite time, every possibility is eventually actualized.

would be nothing existing now since existence begins by something already existing. Thus if at one time nothing existed, it would be *impossible* for anything to have begun existing and even now nothing would exist. Hence, there *must be* something whose existence is not merely possible but necessary. But everything necessary has its necessity either caused or not caused by something else.[5] It is no more possible to go on to infinity in caused necessary beings than with efficient causes. And therefore we cannot but hold there is a being with its own uncaused necessity that, causing a necessity of others, is spoken of as God.

4. Perfection: We experience things as being more or less good, true, and noble etc. But the words "more" and "less", pertaining to nobility or truth per se, are predicated of different kinds of things in terms of degrees of perfection. At the same time, they differently resemble something of maximum perfection. Thus, there is something truest or best or noblest and, consequently, that is uttermost *being* since what are greatest in truth are greatest in *being* (*Metaph.* ii).[6] Now the maximum of any kind of thing is the cause of every member of that genus. So, for any genus, there must also be something that causes their being and every other perfection; and this cause is called God.

5. World's Governance: We see that even natural bodies which lack intelligence do, nevertheless, act for an end.[7] The end is evident in their almost always acting in the same way for the best results. Given the results, it is plain that their actions are not random but rather characterized by design. Now whatever lacks intelligence cannot have a design by which it moves toward an end, unless the end is governed by something else that has intelligence and knowledge; as a knowledgeable

5 *Ibid.*, p. 62, fn. 25: "every necessary being" may refer to unchanging angels in Biblical terms or to bodiless intelligences in Aristotelian terms.

6 *Cf.* Kreeft, *A Shorter Summa*, p. 63, fn. 27. Kreeft notes that degrees of *being* do not mean merely existence "but also essence (*what* a thing [truly] is, its nature), and this latter aspect of being admits of degrees."

7 Is an end of natural bodies related to an equilibrium? Scientists speak of them in physical systems that evolve into biophysical ones whose survival is dependent on their inner mechanics and on their coordination with systems that envelop them. Much has developed along these lines in chaos physics and astrophysics by way of teleological references to "self-organizing systems".

archer governs the end of an arrow by directing it to the target. Thus, there exists a knowledgeable and intelligent being who directs things to natural ends. And this being is called God.

God *via* Negativia and Our Analogical Relation

Though the *Summa* (I, 12, 4) states that it "is impossible for any created creature to see the essence of God by its own natural power", the proofs show that God is *not* moved, *not* created, *not* a mere possible being, *not* imperfect, and *not* governed by anything. These negatives mean that He is *not* in time since only moved movers undergo a "before and after" of existence and change. Also, the negatives mean that He is *not* material since material things change and are imperfect.

Negative knowledge implies a denial of two relationships to God. We do not have a *univocal* one in which we are the same or an *equivocal* one of difference. We have an *analogical* relation of difference and sameness that bears on revelation and reason as well as on modern thought:

> *I. Motion:* As efficient causes, we freely initiate motion and are also caused to move. Thus, we are limitedly both caused and free (in contrast to the extremes of Marx and Nietzsche whose ideas were exploited by feminist ideology).[8]

---

8 In contrast to classical-medieval metaphysics as 'things *after* physics' *(ta meta ta phusika)*, Marx *began* with metaphysics for holding that "truth" in historical epochs is determined dialectically by the movement of matter. Similarly, Nietzsche began with a metaphysics of free will for an unfettered "will-to-power" of Supermen to freely mold reality and truth. Whereas Nietzsche nurtured the incoherent idea of created cultural realities as truth-conditions for relative "truths", where sentence "*S*" may be both true and false in different cultures, Marx held that "truth" is relative to different epochs. Besides the incoherent relativism, he faced the paradox of his own "truth" being caused in his epoch. In holding that *his* own progressive consciousness transcended its epoch, Marx fostered an ironic elitism among some liberals influenced by a radical feminism. When they speak of their *own* "raised consciousness", they assume an elitism attributed to others. The latter are typically members of a male dominating class. For reference to a Marxian "RAISED CONSCIOUSNESS," see C. Kramarae's *A Feminist Dictionary* (Boston: Pandora Press, 1985), p. 106. And for ties of feminism to a Nietzschean relativism, see Kathryn Parsons' "Nietzsche and Moral Change," *Nietzsche*, Ed. by R.C. Solomon (NJ: Anchor Books, 1973), p. 175.

II. Efficient Cause: Though we cannot create things ex nihilo, we are both voluntary and natural causes who freely create things such as art and artifacts of technology. Unhappily, efficient causality was given a so-called "critical expression" by Hume and Kant who transposed it into a causal principle for modern science that admitted only of natural causes.

III. Necessary Being: Since our creations are possible beings dependent on us, we are necessary for their existence and are like the Necessary Being.

IV. Perfection: Though a perennial experience of human nature indicates an impossibility of our obtaining perfection — despite modern ideologies that promise perfection by remaking our nature, we aim for it morally and aesthetically as well as rationally.

V. Governance: In governing order, God orders our analogical order: Christ to Church, Church to state, state to citizen, husband to wife, they to children and so on. In recognizing only *quantitative* analyses, an ideal knowledge of modern science influenced: *(i)* rejection of science for a *qualitative* or normative reasoning from how our nature *is* to how it *ought* to be fulfilled, and *(ii)* acceptance of a radical social-political egalitarianism warned against by Plato and Aristotle alike.

The Proofs & Our Analogical Relationship to God

God's Natural Law and Morality

Traditionally, scientists recognize some coordinating principle of a natural taxonomical order from persons and other less intelligent animals to plants and inorganic phenomena. Modern biophysics may ignore an organizing principle of cosmic Intelligence for its limited methodological purposes, but its nonexististence does not follow.

In viewing God as an Efficient and Final Cause who created every being with "its own... form [*qua* species and]... relation to something else", Nature is viewed in terms of experienced effects that represent an image of divine Intellect and Will (I, 45, 7). This image is manifest in our self-conscious intelligence as well as by the interrelated bio-physical behavior of our immunity systems, by plants and viruses, and by causally-coordinated environmental systems. And because human nature and Nature are good in virtue of a divine creation and end, it is the case both that there are natural virtues which avoid a Naturalistic Fallacy and that their fulfillment by super-natural virtues is understood in the context of various laws.

1. Eternal Law: Thomas states that law is "a dictate of practical reason emanating from the ruler who governs a perfect [complete] community" (I-II, 91,1). Since eternal law involves the whole community of the universe governed by a Divine Reason whose conceptions are "not subject to time but... eternal", all things are subject to this law insofar as their inclinations for proper acts and ends are derived from its "imprint" on them. This "imprint" on rational creatures is unique since, among other things, they provide for others as well as for themselves. *We have natural inclinations for higher acts and ends, which stem from the eternal law, and this imprint on our nature is the natural law.*

2. Natural Law: In reflecting a "light of natural reason" shed by Divine Light,[9] this law is the basis for *natural virtues* of prudence, fortitude, temperance, and justice. The latter are known as practical wisdom, courage, moderation, and fairness. We are not "naturally" virtuous but rather acquire virtuous behavior by practice. Still, our potential practice of the virtues has roots in our natural inclinations which are imprints of eternal law and, thus, tied to our conscience. The latter might be understood as God's word written in our hearts. Whereas Aristotle held that unvirtuous behavior impedes our natural fulfillment and leads to unhappiness, Thomas suggests

9 In rooting *natural reason* in biology, psychology, and philosophy as they apply to a modern controversy, Professor Manfred Hauke, of the University of Augsburg, notes that this application accords with the philosophic, scientific, and theological developments of a Sacred Tradition. Thomas grasped this "Tradition systematically and... [bound] it together with natural reason". See Hauke's *Women in the Priesthood? A Systematic Analysis in the Light of the Order of Creation and Redemption*, Tr. by D. Kipp (San Francisco: Ignatius Press, 1988), p. 445.

that it also violates the sacred and results in spiritual restlessness. When the behavior is unconscionably sustained, it may result in eternal condemnation and separation from God. *Persons not exposed to Scripture, due to cultural or other reasons, may still be judged by God in view of whether they abided by their natural moral light or conscience.*

3. *Human Law:* In most immediately resting on natural law, the moral virtues provide a foundation for *human law.* Though "a law is a dictate of practical reason," says Thomas, "it is from the precepts of the natural law, as from general and indemonstrable principles, that the human reason needs to proceed to the more particular... determinations... called human laws" (I-II, 91, 3). The Principle of Noncontradiction is sometimes held to be an instance of such indemonstrable principles. But the *logically necessary truth* of that principle precludes the derivation from it of any *nontrivial* particular principle. Better candidates for the analogy might be certain principles that underlie science. When Thomas holds that it is impossible for a thing to cause itself, this might be an indemonstrable principle for a more particular law such as "It is impossible for a person to exist when there is no causal agency of male and female parents". And given our moral as well as reproductive nature, we may consider a general indemonstrable principle such as a natural precept of doing good. We may infer from that precept a particular goodness whose virtue is demonstrable by the happiness produced. *We might reason from the virtue of prudence to a more particular human law. This law may specify that a person have one spouse, that a central familial purpose is producing and raising children, and that other laws shall not impede the integrity and well-being of the family as thus understood.*

4. *Divine Law:* Whereas natural and human laws are necessary conditions for ends of earthly goods, the divine law complements those ends for ones of heavenly goods. Natural and human laws are, in a sense, to Old-Testament Law what divine law is to New-Testament Law. As the commandments were fulfilled by the New Law of Love, natural virtues are fulfilled by *supernatural* virtues. These virtues include faith, hope, and love. And they are necessary because: *(i)* there is an ordained end of our eternal happiness, *(ii)* our judgments are uncertain so that reason needs to be guided by revelation, *(iii)* we cannot judge interior motives but only external behavior, and *(iv)* human laws cannot punish every evil or reward every good.

Though the Old Testament acknowledges motives — as when Proverbs warns that God pierces our hearts like a two-edged sword, it emphasizes adherence to law. Christ fulfilled it by transforming the old covenant into a new one of Love. If followed, it guaranteed that we would abide by the old laws in *spirit* as well as in *deed* — as loving our neighbors ensures we do not commit murder in our hearts by hating them, much more by the act. Notwithstanding recent hate-crime laws to which some groups may suppose they are not subject since they do not have the power to institutionalize prejudice, laws and lawyers can only strictly address what is manifest in overt behavior. By contrast: *The divine law bears on all of our behavior since, as we are all one in Christ and by the same token, God judges our inner love or lack of love without respect to race or gender or social status. And on this judgment rests the eternal life and happiness of each individual.*

The Good Political State

The state's goodness is related to Church and family. In this respect, a familial connection to Aristotle is often overlooked. Aristotle emphasizes in his *Politics* that the state depends on families for engendering natural affection between men, women, and children. Moreover, children should be raised ideally in families by both of their natural parents.

Whereas Plato's *Republic* models the state on individuals, *e.g.* the democratic or aristocratic individual, Aristotle argued that families are its basic building blocks and are established "by nature". Thomas extends the thought when he says that "In regard to the whole community, man is perfected... by Matrimony... since it is not only a sacrament but also a function of nature" (III, 65, 1). The state arises from families which, having their origin in natural and human law, are ordained by a divine law of the sacrament of marriage.

The husband is to wife as Christ to Church. As the Church is the body of Christ so that the two are one, the wife is the body of her husband and he is her form so that they are also one. "And Matrimony... touches upon this sacrament [the Eucharist], in so far as it signifies the union of Christ with the Church"(III,65,3). Since the state is formed from families that have sacramental signification, the fate of families and state is inextricably entwined. Accordingly, there is the perennial maxim: AS THE FAMILY GOES, SO GOES THE STATE.

In springing from natural law, the state is a natural institution that codifies human laws. Hence, it participates in the eternal law. Since this law is the Word manifest in the Church in virtue of Christ being its form, the Church is the spiritual form of the state and sets its supernatural end. As the soul is the form of our body, the Church is the spiritual form of the political body. And this body is subordinate to the Church since the Church gives divine authority to the state. At the same time, the state is not an extended Church but autonomous in its everyday governance through laws.

On the one hand, the state does not have laws "for any private benefit, but for the common good of all the citizens"(I-II, 96, 1). An ultimate end of the laws is the spiritual well-being and eternal happiness of citizens. On the other, since happiness is tied to divine law, laws of the state are not "to endure for only a short time, but... for all time by the citizens succeeding one another" (I-II, 96, 1). Without the Church's divine law, other laws do not endure since they are created by imperfect political officials. Their fallible reason is checked by a divine guidance of the Church.

The Modern State and Truth

Thomas held that nontrivial truth is acquired from sense experience. Coupled with our memory of "images" common to experienced particulars, the active intellect affords an induction (*epagoge*) by which the images are grasped as universals. There arise universal principles such as a principle that all things are causally dependent. Knowledge of this dependence, as strange as it may sound, bears profoundly on modern states. Their conflicts, in the first half of the twentieth century alone, have led to the deaths of over 50 million persons — mostly innocent civilians.

How is so much death and suffering related to something as seemingly pedantic as the dependency of all things? The dependency affords an inference to a Cause who is not dependent. In being a voluntary agency and intelligence, this Cause *qua* God renders viable our freedom and reason. And in being manifest in Nature, Nature's God affords inferences from our nature to how it should be fulfilled by political institutions.

The failure of the institutions to fulfill our nature is a source of evil since it frustrates natural pursuits. These pursuits include *(i)* seeking truth about the world in order to work productively and to enhance our understanding, and *(ii)* forming families to raise offspring and to satisfy needs for affection. Importantly, also, the failure of political institutions to fulfill our nature leads to an eventual collapse of the state.

This is a sober thought for fascist and communist states, though the latter survived World War II to witness a collapse later. Communist leader G. Zyuganov admitted that his "state is bankrupt, the president is ill, [and] the government is helpless..."[10] These words, in large measure, reflect the fruits of politicized "truth". Since this "truth" is based on ideological determinations of human nature rather than on how *is* our nature as a truth-condition for scientific truth, the state governed by ideology depends on a cost-ineffective social police (*securitate*) to force unnatural behavior.

This behavior includes loyalty to the state over family and replacing education with indoctrination — under the guise, as Aristotle foresaw, of an egalitarianism in which everyone is friends with everyone (*Politics* 1263). The utopianism, notes A. Codevilla, never had anything "to do with egalitarianism" because the *nomenklaturists* always "lived 'like gods and tsars,'... secure in their own suburbs, eating special foods."[11] But they are not themselves productive. They are parasitic on the labor of citizens whose natural behavior they suppress. The suppression spawns a vicious cycle of unhappiness and declining productivity on which, precisely and paradoxically, the state depends for its power.

---

[10] M. Specter, "Workers Take Over," *The New York Times*, 7 Dec 1996.

[11] A. Codevilla, *The Character of Nations: How Politics Makes and Breaks Prosperity, Family, and Civility* (NY: HarperCollins, 1997), p. 55.

The power to mold or create "truth" is the *sine qua non* of ideology. And philosophers courageous enough to criticize this politicized "truth", as presaged in Plato's *Republic* (489c), will themselves be harassed as the "star-gazing ideologists". Consequently, with no evidence allowed to count against the ideology, which typically mimics natural science, a declining productivity is viewed simply as requiring more unproductive police to enforce infeasible agenda.

Accordingly, the view that there is needed more taxes for a growing unproductive bureaucracy, from the very citizens who are further suppressed, will not be countered by their increasing unhappiness. Mounting social pathologies, caused by sustained unhappiness, will be seen as demanding even more forced behavior with the consequent of a yet greater lack of productivity and need for further force and so on until there is bankruptcy and either anarchy or revolution.

The foregoing indicates how tragic consequences may follow from rejecting the notion that we have a given nature which bears on ethics. The idea that our nature stems from Nature's God, reflected in America's Declaration of Independence, brings us back to Thomas. His naturalistic ethics seems to compete quite well with modern ones that typically view moral *reasons* as "rational" per se or prescribe moral rules with "desirable" *consequences*. Denials of the reasons do not result in self-contradictions. And empirical consequences not only afford mere observation of what is desired but, in ignoring a divine Nature, result either in displacing the desirable from human nature and Nature or in a Naturalistic Fallacy of seeking to deduce the desirable from what is in fact desired.

The desirable is related to Thomas' deduction of a Creator from the fact of the world's existence. That is, in beginning with our experience of existing things in Nature, there was a modal reasoning to Nature's Creator that prevented a despair peculiar to the modern world. The despair is evident from Dostoevsky's claim that without God everything is permissible to an *angst* of notable twentieth-century philosophers over the world's possible nonexistence. Colliding head-on with both the human and natural sciences, the specter of "nothingness" is now considered in light of one of Thomas' proofs of God.

Chapter 7

Profane Despair: Pre-empted by a First Cause?

The words "profane world" denote a world where there is a disregard of what traditional theology holds to be sacred. Above all, the sacred refers to God. Since the time of Kant's "Copernican Revolution" in which metaphysical proofs of a First Cause *qua* God were rejected paradigmatically, the two philosophers who may have most entrenched the rejection are Ludwig Wittgenstein and Martin Heidegger. In view of their pioneering development of anti-metaphysical philosophies and disbelief in a traditional God as Creator or Maker, we may suppose that they had a major role in relegating arguments for such a Maker to archaic syllogistic pedantries; at least among many or most of their mainstream followers who largely dominated both Anglo-American and Continental philosophy. Without denying the stunning contributions of the two philosophers, there needs to be an inquiry into an unaparalleled modern despair (*angst*).[1]

---

[1] For example, see Michael Novak's *The Experience of Nothingness* (NY: Harper & Row, 1971), back cover. Columbia University philosopher Charles Frankel said that the book explores a despair-laden "mood of alienation and disorientation". The eminent philosopher and theologian Bernard Lonergan added that "meaning can be so debased and values so disregarded that... [the] world collapses and there follows the experience of nothingness". Though the word "*angst*" found its way into popular culture in terms of being romantized by rock music and even for naming restaurants, e.g. in Cincinnati, its use to denote an anxiety about "nothingness" has serious philosophical origins peculiar to a secular modern worldview *(Weltanschauung)*.

It is significant that Wittgenstein as well as Heidegger expressed the despair and that this conflicts with a Thomistic optimism rooted in eternal law. In springing from it, natural and human laws should evoke enough virtuous behavior that society, if not having a general ambiance of happiness, does not suffer disoriented despair.

In beginning with considerations of elementary logic that avoid unnecessary symbolism, the inquiry indicates how certain abstract ideas, or their neglect, have far-reaching consequences for social-political institutions as well as for theoretical science. Hence, there is an initial low-level discussion of the first premise in Thomas' second way argument for a First-Cause Creator. This chapter, as opposed to the next several, do not address the argument's soundness but rather its validity after clarifying what is the first premise. Here, a central purpose is to indicate that this premise has a modal status and that, in terms of what is impossible or necessary, this status reflects a viable mode of reasoning about the world.

That there is no stretch of imagination in crediting modal insights to Thomas is indicated by his clear use of modal terms, in arguing from an experienced world, as well as by various scholars who have long noted his cultivation of Aristotelian analyses. Thus for example, Professor E. A. Moody states that his analyses of "Aristotle's *De interpretatione* and *Analytica Posteriora*... [exhibit] some acute discussions of logical questions, especially those pertaining to modal arguments."[2]

Further, after quoting Thomas in several of his works in which one of them states "Bene sequitur, si consequens est impossibile, quod antecedens sit impossibile [*In Metaphysica* 9, 3, 1810]," J. T. Clark, S.J., says that "There is then significant evidence that Thomas knew and acknowledged some theorems of a calculus of modal statements."[3] Indeed, in adding that "There can therefore be no disagreement *in principle* between

---

[2] E. A. Moody, *Studies in Medieval Philosophy, Science, and Logic* (Berkeley: University of California Press, 1975), p. 376.

[3] J. T. Clark, S.J., *Conventional Logic and Modern Logic*, With a Preface by W. V. Quine (Woodstock, MD: Woodstock College Press, Issued by the Secretarial Office of The American Catholic Philosophical Association, Washington 17, D.C.,1952), p. 52.

the Scholastic logic of St. Thomas and a modern logic of statements,"[4] Fr. Clark renders tenable an intention herein to show that Thomas' second-way argument for God may be one that, if not acknowledged surreptitiously by Heidegger and Wittgenstein, is preferable to their related claims about the world which led to either their despair or astonishment.

The immediate discussion first explains why the second way is chosen for examination. Subsequently, the first premise is identified and its modal status established after which the validity of the argument is clarified. Then, there is a comparison of Thomas' reasoning to that of Wittgenstein and Heidegger. Prima facie, if their thoughts about the world and God do not compare favorably to those of Thomas, many pillars of modern philosophy will be in question. This is the case not only because they pioneered the two major twentieth-century philosophies. The philosophies and these pioneering icons, regardless of a revolutionary challenge to modern philosophy itself, were evidently influenced by earlier developments. The developments excluded any "true metaphysics"— in particular, any on the existence of God.

Why the second way?

Among Thomas' five ways, the second way is selected for two basic reasons. First, Heidegger criticizes any appeal to either a divine Cause or Creator by believers. In the attempt by believers to answer "Why are there essents [existents] rather than nothing?", he declares that anyone "for whom the Bible is divine revelation and truth has the answer... even before it is asked: everything..., except God himself, has been created by Him."[5] And although Wittgenstein frequently spoke of a peculiar experience whereby *"I wonder at the existence of the world...',"*[6] he "could *not* understand the conception of a

---

4 *Ibid.*, p. 52.

5 Martin Heidegger, *An Introduction to Metaphysics*, Tr. R. Manheim (NY: Anchor Books, Doubleday & Company, Inc., 1961), p. 6.

6 Norman Malcolm's *Ludwig Wittgenstein: A Memoir*, (NY: Oxford University Press, 1984), p. 59, fn. 4.

Creator".[7] The concept of a divine Cause, much less of a Creator or Maker, was scarcely intelligible to him.

Second, Thomas' second way does not merely go beyond the Cause of motion, in the first way, in virtue of having a divine Cause of existence as well. It involves an efficient causality presupposed in modern science. That is, among the ancient-medieval formal, final, material, and efficient causes, the latter causes alone were transposed from explanatory agencies to a so-called "critical" notion of an invariable succession of events in the modern philosophy of science. This transposition led to a number of troubling epistemic questions.

The questions include how Kant's *synthetic-a priori* causal principle, which has no known truth-value, can be a presupposition of scientific inquiry without undercutting the "truth" obtained by that inquiry. Precisely, seriousness of the dilemma is evidenced by, among other things, the contemporary formulation of a K-K Thesis: "if skepticism is to be avoided [about *K*nowing one *K*nows], the exploitation of... 'causal' regularities in obtaining a posteriori knowledge must not require prior knowledge of those regularities".[8]

Post-Kantian physics and metaphysics are later contrasted to a pre-modern metaphysics of efficient causality. At this point, there is attention to how reasoning may be related to moods. To strengthen the case that the *extraordinary* moods are at odds with an experiencial reasoning is to strengthen the case prima facia that the reasoning reflects an *ordinariness* that makes it more tenable. It is more natural and rational, all things equal, to appeal to the ordinary unless there are extraordinary reasons to the contrary.

Having noted these reasons for selecting Thomas' second way for critiquing central problems of modern philosophy, the first premise needs to be identified and to have its epistemological status clarified.

---

7 *Ibid.*, p. 59.

8 F. Suppe, Ed. *The Structure of Scientific Theories* (Chicago: University of Illinois Press, 1977), p. 722. Based on an international symposium at Chicago Circle, this seminal work expands on dramatic developments in the philosophy of science. The "K-K Thesis" remains problematic.

The Second Way's First Premise

Let the premise be clarified. In being preceded by the sentence "... in efficient causes it is *not possible* to go on to infinity" (I, 2, 3, emphasis added), the last half of the second way states:

> ... if there be no first cause among efficient causes, there will be no ultimate, nor any intermediate [second] cause. But if in efficient causes it is possible to go on to infinity, there will be no first efficient cause, neither will there be an ultimate effect, nor any inter- mediate [second] efficient causes; all of which is plainly false. Therefore it is necessary to admit a first efficient cause, to which everyone gives the name of God.

The passage's first sentence is the first premise: "if there be no first cause..., there will be no ultimate, nor any intermediate [second] causes". But the words "not possible" in the sentence preceding this passage express an impossibility of what cannot occur, given our experience: the impossibility of efficient second causes going on to infinity. Together with a "possible" infinity in the passage's second sentence whereby nothing would exist, the two sentences mean "It is impossible for there to exist second efficient causes when there is no First Cause". Discussed soon, a modal necessity may express the impossibility.

By virtue of the modal impossibility being the outcome of the two sentences which sandwich the first "if..." premise of the argument, the impossibility provides the context for understanding this premise as a modally necessary conditional: "Necessarily if there be no first cause..., there will be no ultimate, nor any any intermediate [second] cause". Taken with the *nonexistent* second causes being "plainly false", there is the conclusion that a First Cause exists to which "everyone gives the name of God":[9]

---

[9] The conclusion does not posit God's necessary existence. Prof. J-P van Bendegem, of Vrije Universiteit Brussel, suggested to me that one of Hume's arguments against any necessary existence, *e.g.* for God in (∀)~NE (x) / E(G) / ∴ NE(G) where for x = G, there is ~NE(G) and ~N(G), ignored one necessary existence -- namely Hume's own for the reasoning! My response was that though his reasoning necessary, it is a necessary condition and raises the question of whether the condition is necessary, *e.g.* Hume's existence. Precisely, its dependence illustrates the starting point of Thomas' reasoning to a First Cause!

FIRST PREMISE:	Necessarily if there is no First Cause, there are no second causes.
SECOND PREMISE:	There are second causes.
THE CONCLUSION:	There is a First Cause.

First Premise's Modal Status

The first premise seems to be understood as "Necessarily if there is no God, there is no world". Prima facie, it is not expressed by a mere material conditional in which the latter is true *if as a matter of fact* it is not the case that "there is no God" is true and "there is no world" is false. Evidently, Thomas holds that it is *impossible* for the falsity to obtain when the antecedent is true. By contrast, for instance, astrophysicists might assert that if there had been no quantum fluctuation of a primordial black hole in terms of various universe and multiverse cosmologies, the universe would not exist.[10] But whereas the scientists would presumably acknowledge the possibility that there exists the present spacetime universe even if a particular "Bang" did not occur, as understood in a given cosmology, Thomas' conditional is connected prima facie to the idea that it is necessarily the case that without a First Cause *qua* God the universe could not exist.

Necessity and impossibility

Thomas' thought seems to be that *necessarily* if there is no God, there is no world, where our experience of the world leads ineluctably to this conditional. That is, for God's nonexistence to permit the valid inference to there being no world, it must be *impossible* for "there is no world" to be false when "there is no God" is true. "There is no God" entails "there is no world" *if and only if* "There is no God, therefore there is no world" is a valid inference: To assert that "there is no God" entails "there is no world" is to assert that the conditional "If there is no God, then

10 See physicist V. J. Stenger's "The Face of Chaos," *Free Inquiry* 13 (1993) 13-14. The black hole prior to the fluctuation *qua* Bang would have maximum entropy or peak disorder as well as be a virtual "*nothingness*" devoid of space-time, structure, and governing laws of physics. On the peak-entropy scenario, the formation of any given universe would be unpredictable.

there is no world" is not merely true but necessarily true. The manner in which Thomas reasons to the premise will be expanded on shortly. At this point, the foregoing indicates that his reasoning falls within the domain of modal logic. The latter is a logic which, among other things, addresses what is experienced to be necessary or impossible about either the nature of things or the world in general. When Thomas speaks of efficient causes in terms of "ultimate" and "intermediate causes", he means either any given individual second cause or causes collectively as the world.

Logical and modal necessity

Certainly, Thomas' reasoning admits of distinguishing modal necessities from purely logical ones in terms of such things as the Principle of Noncontradiction or conclusions that follow premises with logical necessity. But the premises of his arguments tend to stem from a common-sense (*sensus-communis*) experience, reminiscent of Aristotle's "experience of ages", by which they may often be understood to have a modally necessary truth. The truth conceived by Thomas was, surely, influenced by Aristotle. The breadth of Aristotle's reasoning may be why some Anglo-American philosophers, despite an anti-metaphysical indebtedness to logical positivism, credit Aristotle as presaging analytic philosophy.[11]

The ordinary-language philosophy of Wittgenstein and phenomenology of Heidegger will soon be related to Thomas' reasoning. Consider at this point how a modal reasoning of Thomas and Aristotle are connected in terms of Aristotle's frequent attention to modalities of propositions. The latter range from the proposition "It is necessary indeed, if animal follows man, that it should follow all these also [subjects of the predicate 'man']", in the *Prior Analytics,* to the sort "*A* cannot inhere in *B* when *B* inheres in *C*, with the resulting inference that *A* inheres in *C*, and this is a known and admitted impossibility" in the *Posterior Analytics.* However, again, a modal language of efficient causality would be different for

---

[11] For example, see Trundle, *Ancient Greek Philosophy,* p. 8, for reference to Aristotle being credited as a central forerunner to, what is variously called, "linguistic analysis" or "analytic philosophy".

Aristotle and Thomas. Whereas Aristotle's causality involved the cause of change — the "cause of form informing matter", Thomas' second way included this cause and a cause of the world. Consider what is at stake.

A reminder of what is at stake

Thomas' second-way argument, aside from modal terminology, is a *modus tollens:* $\sim p \longrightarrow \sim q \;/q\; //p$, where the symbol " / " separates premises and " // " the premises from conclusion. Beyond indicating how the first premise makes the argument a modal analogue to the valid form --- its soundness soon noted in terms of the premises' truth, there needs to be a reminder of what is at stake in the present discussion. At stake is the epistemic status of the first premise, which is important because it is the controversial one, and modern concepts of "truth" in a context of general norms of reasoning.

The present claim that the first premise has the status of a modal truth is significant if modern assessments have disregarded the status by assuming only logical and empirical truth. Also, the first premise stems from an experiential reasoning. If this reasoning contrasts to angst-laden thoughts that ignore ordinary experience, there may be more grounds for questioning the thoughts about a possible nonexistence of things than an ordinary acceptance of them. The acceptance may be explained by the notion that there never was "nothing" but always "something"— if not the world, an uncaused Cause or God. Such a God need not be explicitly acknowledged by philosophers to submit that some sort of implicit reasoning, resting on ordinary experience, lies behind their lack of angst.

The grounds for preferring Thomas' reasoning may hold quite apart from strictly knowing the truth of his premise. The grounds for selecting alternative sorts of reasoning may be analogous to modern theory-choice in science. Apart from issues of scientific truth and all things equal, scientists generally accept the rationality of choosing between competing theories when one conflicts with *(i)* an ordinary experience of the world, as often occurs in the human sciences, or *(ii)* naked-eye or instrument-aided predictions of observed phenomena in the natural sciences. Even if the natural sciences only sometimes exploit ordinary experience by manipulating instruments at hand, these sciences depend on what may be called an ordinary "Cartesian reality". A dependence on it for confirming observed predictions of quantum or relativistic theories is paradoxical when, as sometimes occurs, the theories are alleged to reveal an illusiveness of the reality.

A case for Thomas' first premise may start with how it reflects such a reality in terms of an "experience of ages". To strengthen the case for an ordinariness of relating the Cause to the world may be to illustrate a disingenuous of their disassociation — especially when extraordinary moods are evoked. Accordingly, consider ways in which the premise may echo long-standing modes of reasoning that seem more viable than the modern ones in question.

Ordinariness of the Premise

Despite their lack of angst, many modern philosophers might object that since the ancient Greeks did not generally relate a divine Cause to the world, they do not evidence an ordinariness of relating the world to a supreme Being. It may be argued that the Greeks did not believe in such a Being in any Thomistic sense but rather, at best, in mythological gods or in some sort of amorphous pantheistic "god".

Socrates is a good place to start in making a case for the ordinariness, of Thomas' reasoning, in response to the objection that ancient philosophy does not link the world to a Cause *qua* God. The entire history of Western philosophy prior to Socrates is named "Pre-Socratic Philosophy". Many undergraduate programs treat one semester on ancient and medieval philosophy as an equal, or even outmoded, counterpart to another on modern and contemporary philosophy. And many contemporary philosophers such as F. K. Waismann may say that there has been philosophical progress by virtue of an analytic method that "leads us to the view that... [a history of philosophical] questions... rest upon misunderstandings".[12]

However, in addition to the fact of Heidegger holding that Heraclitus and Parmenides were far more profound than any

12 Friedrich Waismann, *Principles of Linguistic Philosophy*, Ed. by R. Harre (NY: St. Martin's Press, 1968), p. 5.

contemporary philosopher, ancient Greek philosophy did not merely span a greater historical period than the history of modern and contemporary philosophy put together. In contrast to most of the latter philosophy, it did not splinter our lives by either ignoring or denying entire domains of moral, spiritual, and aesthetic experience.

The words "Pre-Socratic Philosophy" underscore Socrates' influence and his assimilation of most of the earlier essential insights. Thus the objection ignores his intriguing assertions, in the *Apology* 23a, that "real wisdom is the property of God" and that "human wisdom has little or no value." The assertions are strikingly similar to St. Paul's assertions in I Cor. 1:20: "Where is the wise man? Where is the scholar? Where is the philosopher of this age? Has not God made foolish the wisdom of the world?" Moreover, the objection ignores a continuity of developments from the Pre-Socratics to Aristotle. Religio-scientific developments, in regard to the observable Many and inferred One did, according to Aristotle, lead to his view of the One as a divine ultimate Cause.

Some modern philosophers may grumble that Aristotle's history of philosophy, in the first few books of his *Metaphysics*, was self-serving. But he was closer to the philosophy than they and said in his theology that a search for this Cause was one for God qua Wisdom. The search was not for mere cleverness (*sophis*). The Sophists' relativism and Atomists' science-oriented materialism, which rejected any gods such as Thales' "gods in things", may be associated euphemistically with the Greek Enlightenment. However, beyond the fact that this Enlightenment no more ended religio-philosophical developments than the modern Enlightenment, Socrates as well as Plato and Aristotle noted the incoherence of a Sophistic-Atomistic metaphysics. The latter was concealed under a rubric of either "physics" or "rhetoric". The concealment is evident in a host of works that include the *Phaedo* 98-9, *Republic* 344c, and *On Sophistical Refutations* 164a-184b.

Finally, in addition to Socrates speaking of God as noted, Plato spoke of a Demiurgos who shaped the world as a sculptor does marble and Aristotle of God as resembling the supreme Being in Homer's *Iliad*. Since the "world refuses to be governed badly, says Aristotle in the *Metaphysics* 1064, "one ruler let there be." Even when he views God as a Form of the world's substance — which he was led to do by his criticism of Plato's Form-particular distinction, he understands God as more than an epistemological principle.

This discoverer of formal logic spoke ontologically of God, in the *Metaphysics* 1072a, as a "living thing" who "thinks of that which is most... precious", who produces motion "as being loved", and who "life also belongs to." A supreme-Being thesis is not being glibly advanced for Aristotle. And an Aristotelian thesis of God as a mere Form of the world's substance is not being suggested for Thomas. The above references are reminders that, besides some linguistic ambiguity in Aristotle's otherwise precise language, he provided fertile seeds for Thomas' First-Mover and First-Cause arguments.

Hopefully, the notion has been strengthened that there is historical evidence for an ordinariness and naturalness of reasoning to a divine Cause. Accordingly, we may reemphasize the difficulty of explaining perennial references to the Cause when it is not fundamental to an experienced world. An ordinariness of reasoning to the Cause is complemented by an inferential direction, which is now considered, of reasoning to an impossibility of the world's existence when there is no causal Creator *qua* God. This inferential direction seems in contrast to a direction of Heidegger and Wittgenstein. Their disregard of God evidently involved a possibility per se of the world's nonexistence. Whereas they begin with a modal possibility per se, Thomas infers an impossibility from experience.

Chapter 8

The Cause from Modally Relevant Experience

Two points need to be made before considering Thomas' inferential direction of beginning with the experienced world for inferences to a modal impossibility. First, today, the possibility of the world's nonexistence would tend to be understood in terms of a "contingent truth". It has been said that whereas the Jesuit looks at a blade of grass and proclaims God's existence, the modern philosopher declares that existence is simply contingent.[1] According to a modern philosopher's received view of 'truth', empirical truth ascribed to statements about the existence of things is contingent on what either might or might not possibly *be* — as opposed to any truth about *being* being logically necessary.

Later, more shall be said on how a restrictive modern conception of there being only logically necessary and empirical "truth" may result in underestimating Thomas' claims. For now, it is stressed that the claim that something either exists or is a certain way, to which empirical truth may be ascribed, is not to be confused with a claim that whatever *is* might not possibly

---

1 Professor Terry Pence is gratefully acknowledged for providing the "Jesuit-philosopher" example. The example may later find modal expression by the following:

$$(S_o \Rightarrow F_o) \wedge S_o \;/\therefore\; F_o$$

Here, "S_o" denotes second causes, such as a blade of grass, and "F_o" designates a First Cause that is understood as God.

be to which the truth does not seem properly ascribed. Though the claim does evidently have a modal status and not an ordinary empirical one, the question ensues of whether one fruitfully begins — if not how one is to begin — with substantive existential "possibilities" or "impossibilities" apart from experience as opposed to starting with experience for reasoning to them.

Second, supposing the existence of a God *qua* First Cause, the question might arise of how it could diminish an *angst* over the world's possible nonexistence. There might an objection that the impossibility of there being a world when there is no Cause does not, in principle, address an *angst* over a specter of "nothingness". The world's *necessary* existence is not being claimed but rather only its impossible existence without the Cause. One response is that the second way may specify that the Cause is not dependent on other causes for its existence. Thus, without the world there would not be "nothing" but rather the Cause. Though its existence may raise a further question of why it exists, the existence contrasts to possibly nonexisting things which led to the various moods. Moreover, while the premise's truth has not yet been established, these considerations are not only enough to warrant a clarification of inferential directions, in which one is based on modal possibilities and the other on experience, but indicate prima facie that Thomas' experienced-rooted direction is preferable to that of either Heidegger or Wittgenstein.

Beginning with Modal Possibilities

Let us consider Wittgenstein after Heidegger with respect to general modal directions of reasoning. Thus, to begin with, Heidegger treats nonexistence as an attribute wherein a "large thing" having the attribute of "lying here" just as surely has the "attribute of potentially not lying here and not being so large".[2] The potentiality of *not being* "such and so" is merely a less "extreme possibility of nonbeing...,"[3] and this extreme

---

2 Heidegger, *An Introduction to Metaphysics*, p. 25.

3 *Ibid.*

possibility holds not merely for given existents but for the nonbeing of their "totality". The totality is understood as the world. We need to keep in mind that the world and any given thing in it do not have the *potential attribute* of nonbeing but rather already the *attribute* of potential nonbeing. Whereas Heidegger reasons overtly from a modal possibility of things *not being* to a possible *nonbeing* of their totality, Thomas reasons from experience — of *being* as being dependent — to a modal impossibility of there being experienced things when there is not something which is not dependent. This point is soon supported by his own statements.

And ironically as soon discussed, Heidegger may have begun covertly in the fashion of Thomas with things *being dependent* for inferring possibilities of *nonbeing*. Let it first be shown that if he began with mere possibilities of *nonbeing*, as he evidently thought he did, a consideration of inferential directions indicates the merits of Thomas' approach. Again, the approach is also relevant to Wittgenstein . Herein, an inquiry into their approaches will consider how beginning with various modal claims either lead or do not lead to definitive inferences about existence. By also regarding necessity and impossibility, greater light is shed on peculiar limitations of reasoning from a modal possibility.

1. Necessity: For example, we may simply suppose that a person named Boethius necessarily exists. What is inferred from the necessity of his existence is that he does exist. Modal necessity in this sense means that from our assertion that something must either exist or be a certain way, e.g. large or small, it follows that it exists or is that way. The modal supposition and inference may be unusual and unsubstantive, but the reasoning can obtain.

2. Impossibility: Similarly, our supposition that Boethius' existence is impossible means he cannot exist. So, what may be inferred from this impossibility is that he does not exist. This sort of modal impossibility illustrates that from our assertion that something cannot either exist or be a certain way, it follows that it does not exist or is not that way.

3. Possibility: Finally, in supposing Boethius' possible existence, there is no implication for either his existence or nonexistence per se. To say that he possibly exists leads simply to the inference that he might or might not exist, but not to the inference that he does or does not. Moreover, beginning with a claim that he exists *and* might possibly not exist does not lead to any significantly stronger inference. As emphasized momentarily, the claim cannot be tenably bifurcated from the mere possibility of his nonexistence.

Though claims about what is necessary or impossible yield inferences to what do or do not exist (or are, or are not, a certain way), claims about *possibilities* do not permit definitive inferences. Now, extraordinary moods that are evoked over the possibility of there being a world might seem to be starting with tacit nonmodal claims that there is a world.

Certainly, one might object that the question of why there is a world presupposes that "there is a world." But in addition to equally presupposing that "there need not be a world," a rule of sentential logic specifies that the truth of their conjunction requires that they both be true. And a further objection that by the rule of *simplification* we can infer that there is a world, disregards that by *addition* we could then infer "Either there is a world or there need not be a world" where one or the other or both disjuncts could be true. The latter is the weakest inference which is not logically equivalent to either the simplification or the conjunction in which both conjuncts need be true:

WHY IS THERE A WORLD?

1. *Presupposition:*	There is a world and there need not be one.
2. *Simplification:*	There is a world.
3. *Addition:*	There is a world or there need not be one.

The point about these trickle-down inferences is that the first would generally be taken to be stronger than the second of simplification and third of addition. Thus on the the strongest immediate inference, the question "Why is there a world?" cannot be bifurcated from a modal possibility of there being no world — of there being "nothing".

Surely, an analytic approach to philosophy may suggest that the presupposed conjunctive sentence is nonsense. The *sense* of the question might be belied by its not being answerable in an ordinary way. "Think...," said Wittgenstein in a complimentary article on Heidegger, "of the astonishment that anything exists. This astonishment cannot be expressed... [by] a question, and there is no answer to it."[4]

Yet Wittgenstein's remarks, given his astonishment, do not obviate its implicit modal status. Nor did his uneasiness with unanswerable questions, dramatized by his Anglo-American followers, lead him to disavow the importance of such putative "nonsense".[5] Wittgenstein appealed to St. Augustine, whose writings he revered, when he quoted the Saint as saying "What, you wretch, so you want to avoid talking nonsense? Talk some nonsense, it makes no difference!" (*"et vae tacentibus de te, quoniam loquacaes muti sunt"*).[6]

Notwithstanding their vastly different philosophical approaches, Wittgenstein and Heidegger are not starting with claims about how is the world. They are beginning with either implicit or explicit claims which specify that whatever *is* the case need *not be*. The modal nature of this possibility, over which there is astonishment or despair, cannot in principle result in definitive inferences about the world, much less in

---

4 L. Wittgenstein, "On Heidegger on Being and Dread," *Heidegger & Modern Philosophy*, Ed. by M. Murray (New Haven: Yale University Press, 1978), p. 80. Importantly, points made about Wittgenstein herein are not undercut, prima facie, by recent controversies surrounding his *Nachlass* which might eventually liberalize various interpretations. See *The Cambridge Companion to Wittgenstein*, Ed. by H. Sluga and D.G. Stern (NY: Cambridge University Press, 1996), *e.g.* pp. 442-476.

5 *Ibid.*, p. 82. Editor's note. This little-known article was preserved by F. Waismann and first published in the *Philosophical Review* (Jan 1965). But it was a "sanitized" version in which Heidegger's name was deleted in order to make it "acceptable" to Wittgenstein's followers. Since the time of Carnap's "The Overcoming of Metaphysics" (1931) and Ayer's *Language, Truth, and Logic* (1936), Heidegger's thought was held to be a "paradigm of the worst".

6 Wittgenstein, "On Heidegger on Being and Dread," p. 81. The editor notes that "According to [B.F.] McGuinness, this was a favorite quotation of Wittgenstein's. Though the exact passage has not been found, he suggests the *Confessions* I, iv...".

modal truths that have a stronger epistemic status than ordinary empirical truth. In the next chapter, there is a defense of the stronger truths in terms of Wittgenstein's own investigations. With limited reference to them here, Thomas' inferential direction is now examined.

Modalities from Experience

In beginning with experienced things, we can appreciate a more tenable inferential direction. The direction is one in which Thomas infers his modally necessary premise. Since elementary modal necessities and impossibilities have been clarified, we may examine how they are typically taken with everyday experiential claims. The beginning of Thomas' second way, elaborated on shortly, states that "In the world of sense... [there] is no case known (neither is it, indeed possible) in which a thing is... the cause of itself".

Let us recall the oddity of not starting with the "world of sense" but with existence claims of modal necessities. Though the necessities may lead to inferences that persons "do exist", they ignore whether the existence is fact. And notwithstanding facts that pertain to biological technologies such as cloning, considerations of our reproductive nature may lead to modal insights: "Necessarily if there are not male and female parents, human offspring are not produced" in certain contexts. Taken with our experience of Boethius' existence in either a past or present-day ordinary context, the impossibility he exists without both parents may be inferred.

Though everyday thinking is sometimes difficult to explain by logic, this example may explain why we would not ordinarily question whether a miracle occurred if it was said that Boethius had no parents. We would naturally assume circumstances in which, say, they were no longer alive.

The assumption's significance is reflected in the religious belief that Jesus had only one biological parent. As disbelievers do not appeal to the logical impossibility of his divine conception but to its conflict with our reproductive nature, they do not hold that this nature is understood in terms of a logically necessary truth. If there is a truth about it, disbelievers and believers alike suppose it is weaker than a logically necessary truth but stronger than an empirical one; say, that someone exists. Otherwise, a denial of the possibility of Jesus' virgin birth by disbelievers would be inexplicable. And importantly, after suggesting that someone's birth without two parents is *more* than merely improbable, aside from biotechnical interventions, Wittgenstein notes that "*Very* intelligent and well-educated people believe in the story of creation in the Bible...".[7]

The Biblical account of the world's creation is, of course, no more astonishing than the created existence of Jesus. The account of Jesus needs to be briefly contrasted to scientific reasoning. In a recent medical inquiry about a boy whose conception lacked normal fertilization — in circumstances apart from any artificial processes, a researcher did not speak of a logical impossibility of his conception in a female parent's egg without a male parent's sperm. The researcher announced that "This was not an immaculate conception... Studies in other mammals suggest that *cannot* happen."[8] Here, the word "cannot" clearly denotes a biological impossibility. It is not terminology about mere empirical facts but about what is supposed for epistemic parameters of their possibilities. More will be noted about this later.

Now recalled is that the question of Wittgenstein and Heidegger, of why there is a world, presupposes experience of it. The presupposition implicitly reveals a need for experience to which Thomas explicitly appeals. But the question also presupposes the world's possible nonexistence. And this possibility not only excludes any inference about the world but modal necessities and impossibilities by which sense is made of modal claims commonly acknowledged. While the acknowledgment is expanded on later, the ones considered thus far are

---

7 L. Wittgenstein, *On Certainty*, Tr. D. Paul & G.E.M. Anscombe (London: Harper & Row Publishers, 1969, p. 43e #335-6.

8 Malcolm Ritter, "Egg Didn't Wait," *The Associated Press* 29 September 1995. Emphasis added. In Reference to his article in *Nature Genetics* (October 1995), Edinburgh University researcher D. Bonthro suggested possibilities consistent with male fertilization. I am grateful to Professor J. Roland Ramirez for distinguishing the Immaculate conception of Mary from Jesus' virgin birth.

sufficient for evidencing the preferableness of Thomas inferential direction. They also bear on the role of disbelief:

Specifically, there is the modal necessity about our reproductive nature a denial of which would be deemed physically impossible by disbelievers. The impossibility is inferred from experience to which Thomas does not appeal for what is known only by faith. At the same time, faith is a short step from Thomas' inference to a Creator on the very basis of a dependence that disbelievers embrace. This specter puts a novel spin on the Biblical notion that God makes foolish a wisdom of the world. He uses simple things to astonish sophisticated persons who, in the cases at hand, were often members of the Greek and modern Enlightenments. They tended to be influenced by science in disavowing not only a rationality of religion and a Creator but the consistency of Creation and Evolution.

Summary: Reasoning to the Premise

The foregoing accords with Thomas' reasoning to the syllogism's first premise. The reasoning goes from experience to a modal claim that in "causes it is not possible to go on to infinity". These words essentially introduces the premise that there cannot be a world of second causes without a "first efficient cause". Keeping in mind that this cause ultimately causes both existence and change, the reasoning to the modal first premise, in the second way, is as follows:

> The second way is from... efficient cause. In the world of sense... there is an order of efficient causes. There is no case known (neither is it, indeed, possible) in which a thing is found to be the efficient cause of itself; for so it would be prior to itself, which is impossible. Now in efficient causes it is not possible to go on to infinity...

In starting with an experienced order of second efficient causes, Thomas does not merely appeal to our knowledge of particular cases (*e.g.* "no known cases"). He infers several modal impossibilities.

First, he infers the impossibility of a thing being its own cause. He may seem to say that its being prior to itself is a logical impossibility since it would have to exist and not exist at a prior time. But intrinsic to the context is a modal inference from experience. Experiencially, we grasp that it is in a thing's primordial nature to be caused by some prior thing — as evident by such things as our reproductive nature. And thus we infer various modal necessities and impossibilities.

For instance, in asserting that "It is impossible for a thing to be uncaused", there is no appeal to logically impossibility but rather to a modal impossibility in light of the experienced world. If Thomas' notion of the impossibility of a thing being "prior to itself" does not consist in modalities inferred from experience, he could be construed with Parmenides to rely on mere reason: Change would involve *being* coming to *be*. However, *being* would have to come from either *being* or *nonbeing*: *Being* cannot come from *being*, since it already *is*, nor from *nonbeing* since there is nothing to come from.

Second, the statement "it is not possible for second causes to go on to infinity" is related to an inference from second causes making up the world to the world's dependence (I, 46, I). It is odd that many logic textbooks do not address the fact that this no more commits a fallacy of "composition" than arguing that since every atom has mass, a phenomenon composed of atoms has mass. Unless the phenomenon we call the universe is causally dependent, astronomy as a study *inter alia* of its cause would be quite unintelligible.

A question-begging explanation might involve a hypothesis of an uncaused cosmos within which causality pertains only to multi-universes:[9] Universes within the cosmos would go on to an infinity of collapses and refluctuations that cause their destructions and creations. But given an experienced nature of phenomena and aside from logicians who may speak of *logical* possibilities, the question remains of whether any scientist would allow for a *modal* possibility of there being no cause of this ultimate cosmos. To accept its being a phenomenon per se would be to accept, on the face of it, either a finite or infinite causally-dependent cosmic series that renders coherent our scientific inquiry.

9 Cf. O. Gingerich's "Space, Time, and Beyond," *Gross Memorial Lecture* (IN: Valparaiso University Press, 1993), p. 16. Though this explanation is not explicitly submitted, the specter of an uncaused causal series is clearly raised.

Given the fundamentally dependent nature of both phenomena in the cosmos and cosmos itself — presupposed by science from astrophysics and physics to biology and psychology, either a finite or infinite series of cosmic phenomena and phenomena within them would still *be* dependent: By its rudimentary nature, the series would not be sufficient to cause its own existence and be dependent on an uncaused First Cause sufficient to cause its existence. However, this Cause cannot be a mere phenomenal being such as human beings who consist of dependent bodies and minds, *i.e.* of mere *potent* natural and voluntary rational capacities. It would have to be an *omnipotent* voluntary rational Cause.

That is, based on our own incontrovertible self-conscious experience of ourselves as natural and voluntary second causes, only causes which are voluntary may not have a natural cause per se even if *our* existence is causally dependent on both sorts of causes, *e.g.* self-consciously chosen sexual behavior. And a rationality coupled to a voluntary First Cause is necessary prima facie to account for an ordered cosmos.

In short, we discover truth wherein, for example, true or approximately true quantum and relativistic theories coordinate an ordered cosmos to a disordered causally dependent cause, *e.g.* a black hole with maximum disorder. And we are ourselves rational second causes who may voluntarily reason from dependent things. These include modalities as second-causes caused by our thoughts such as "Necessarily if there is no First Cause, there are no second causes". And in presupposing the causes, this necessity underscores that their causal dependence is necessary for scientific inquiry. Inquiries in general physics relate to metaphysico-modalities expressed quintessentially by this modal necessary which, taken with "There are second causes", yields "There is a First Cause".

Modern philosophers may question the Cause's identity with a Judeo-Christian God. Yet as unsympathetic as Wittgenstein and Heidegger were with a science paradigm of knowledge, there is little doubt *they* viewed the Cause as a Creator who amounts to the God. Despite this, Thomas distinguishes this God from the Creator. Heidegger says that "Anyone for whom the Bible is divine revelation and truth has the answer to... 'Why there are essents rather than nothing?' even before it is asked..."[10] But his starting with the question involved a modal possibility from which no answer could be forthcoming. Also, of course, Thomas begins with experience and not with revelation.

And while Wittgenstein declared that the existence of God "isn't a question of *proving* anything!,"[11] his assertion "Not *how* the world is, is the mystical, but that it *is*"[12] reflected a mysticalness different from revelation. His reference to the mystical revealed his deep respect for religious belief. But he regarded much of it as being "based on qualities of character and will that he himself did not possess".[13]

Unhappily, these points bear on modern philosophers who too hastily disparage any rational content of Scripture. In appealing to St. Paul's assertion that "For the original Christian faith philosophy is foolishness,"[14] Heidegger may reveal a foolishness of philosophers who deny that philosophy and faith are complementary. In view of Wittgenstein's own philosophy, they might be culpable for their failure to acknowledge what is being said and what can only be shown. A little-known paper on Wittgenstein's appearance before the Cornell University philosophy club in the late 1940s contains intriguing remarks on him by former graduate students:[15]

> I thought at the time I'd undergone a conversion, but what I'd received... was a philosophy shown, not... argued. Wittgenstein had uttered what he felt could be... [He had] a method, and the moral and aesthetic passion of a mind in love. How pale seems Sartre's *engagement* against the deep and fiery colors of that

10 Heidegger, *An Introduction to Metaphysics*, p.6.

11 Malcolm, *Ludwig Wittgenstein: A Memoir*, p. 59.

12 L. Wittgenstein, *Tractatus Logico-Philosophicus*, Tr. D. F. Pears and B. F. McGuinness (London: Routledge & Kegan Paul, 1961), #6.44.

13 Malcolm, *Ludwig Wittgenstein: A Memoir*, p. 60.

14 Heidegger, *An Introduction to Metaphysics*, p. 6 (I Cor. 20).

15 Professor William Gass, "Wittgenstein -- A Man and a Half," *Witt-genstein's Appearance Before the Cornell Philosophy Club*, pp. 4-7. I am beholden to Professor Emeritus John Nelson, University of Colorado at Boulder, for a copy of this memoir in about 1980. He worked on his PhD under Norman Malcolm and contributed to it.

> purely saintly involvement. It now seems inevitable that the *Tractatus* should have stressed, so much, the difference between what can be *said* (and anything... said can be said clearly), and what can only be *shown*..."

In traditional Christianity, especially in Augustine's existential thought, there is emphasis on what can only be *shown* by personal exemplars. Christ is the quintessential Exemplar. Here, nothing is contrary to Thomas' limitations on what can be *said*. But the latter has more significance when modalities are inferred from experience. Despite Wittgenstein's silence about experiential modalities, the above indicates how closely he, if not Heidegger, came to faith. The "ideas of Divine judgment... and redemption had some intelligibility for him," notes Malcolm, as being related to "a sense of the helplessness of human beings to make themselves better".[16]

Contrast to Modern Metaphysics

Wittgenstein's notion of a helplessness in the human condition may be more reminiscent of Augustine, whom he revered, than Thomas. However, both theologians held that an experiential reasoning was necessary for even profound thinkers who have extraordinary integrity. Thomas addressed limitations of the integrity in terms of a need for sacred doctrine.

At the same time, he noted the doctrine uses "reason, not, indeed, to prove faith...," which Wittgenstein agreed is not a matter of proof, "but to make clear other things" (I, 1, 8). Thomas viewed philosophy as an activity for the clarity in virtue of I Corinthians 10 where St. Paul advises bringing "into captivity every understanding into the obedience of Christ". The understanding found expression in natural reason.

Explanations of why natural reason came to be rejected were previously noted. The following considers consquences of confusing a speculative metaphysics with a metaphysico-modal reasoning. This latter reasoning led to a modal truth "All events have natural and voluntary causes" which *wedded* causal agencies that were *divorced* by Kant. A metaphysics of

---

16 Malcolm, *Ludwig Wittgenstein: A Memoir*, p. 59.

determinism and freedom not only had no truth-value but freedom was a mere possible reality.

Moreover, Kant's metaphysics, as presuppositions of moral and scientific inquiries, were *insufficient* for any truth. On the one hand, moral truth-claims presuppose freedom but it does not presuppose any morality — grasped by Nietzsche when he celebrated Kant's new "Copernican freedom" to create truth.[17] On the other, an unqualified determinism undercuts the truth of even the natural sciences since truth-claims would themselves be caused. A causally determined reality would, in effect, be the truth-condition for "truth" so that contradictory claims of scientists would incoherently both be true.

By contrast, causes pertaining to the medieval principle are related modally to a voluntary First Cause. Since this Cause renders good a created Nature, scientific descriptions of our nature are the basis for moral prescriptions of how it ought to be fulfilled. And this notion of fulfilling our nature is not merely common sense but excludes a Naturalistic Fallacy because prescriptions are implicit in descriptions. The "ought", so to speak, is implicit in the "is".

When philosophers stress that "truth" about how we *ought* to behave cannot be deduced from descriptions, they seem to studiously ignore that it is not ascribed to descriptions of scientific theories either: Material implication incurs the logical possibility that the theories may be false when the predictions are true. And there is no known truth of a presupposed causal principle that admits only of natural causes and is rejected because it is merely metaphysical.

Anti-Metaphysics and Political Truth

In failing to answer "Whence comes truth if not from metaphysics?," modern anti-metaphysical philosophies nurtured scientific and moral truth-claims based on politics. The Aristotelian-medieval tradition underscored that to claim political agenda are good is to be able to say what is "good". The "good" included fulfilling our nature on the basis of

---

17 *Cf.* for example, Friedrich Nietzsche, *The Will to Power*, Tr. by W. Kaufmann (NY: Random House, 1968), p. 416.

biophysico-psychological theories that are true. Modern secular philosophies, by contrast, did not merely fail to articulate how objective truth is ascribed to scientific theories. They failed to avoid metaphysical presuppositions, such as a causal principle, that were also typically related to metaphysical frameworks.

The frameworks of Whitehead and Einstein, for example, were the "real-possibilities" and "block-universe" worldviews. The latter stemmed from quantum and strictly deteriministic interpretations of the causal principle. Though without truth, they were assumed to truly bear on human nature and moral norms such as retributive and rehabilitative views of punishment.[18] Given that scientific truth presupposes truth-valueless interpretations of a principle related to such worldviews, what was to prevent views of pseudo-scientific ideologies without truth? Of transposing the notion of "truth", which the views do not have but seek to mold, into the idea of "correctness"? A "political correctness" in universities of the West, that rings of an epistemic conformity — if not of a totalitarian lock-step thinking, raises a specter of ideology in the societies to which turned the post-communist East for models of reform.

There can be little doubt that elements of Marxism and fascism, in which political power has the ultimate say about human nature and Nature, have been exploited in open societies since the cold war for assuaging conflicts they inflame.[19] One need only note Gustav Bergmenn and Nicholas Rescher to entertain plausible ties of the inflammation to philosophy. Bergmenn's contributions to Einstein's theory, relegated initially by the Nazis to "Jewish physics", ended with their ascension to power. Tellingly, his shift away from the anti-metaphysics of positivism "with his work in *metaphysics*," occurred after the war at about the time he stated: "The ideal of an ideology-free society is... to be desired, if for no other

18 D. Sherburne, "Responsibility, Punishment, and Whitehead's Theory of Self,"*Alfred North Whitehead: Essays on His Philosophy*, Ed. by G. L. Kline (NJ: Prentice-Hall, 1963), pp. 179-188.

19 See R. Trundle's "Cold-War ideology: an apologetics for global conflict," *Res Publica: Leuven Politologisch Instituut* 37 (1996) 61-84.

reason than the humanity, the intelligence, and the courage it takes to bear life without... ideological illusion."[20]

In the *American Philosophical Quarterly,* Rescher suggests how there may be ties of ideological illusion to a self-refuting verification principle.[21] In addition to an issue of intellectual honesty, its zealous proselytization led to an ideology-like silencing of alternative views and to dictatorial behavior in hiring faculty.[22] Also, in being used to keep a "meaningless" metaphysics out of physics and in virtue of physics being ideal knowledge, the principle nurtured scientific *Weltanschuungen* (worldviews) that inspired equally fervent political opinions. Rescher quotes a broad spectrum of eminent philosophers who either politicized truth or rationalized terrorism. He laments that many philosophers are "just as captivated by political fashions and frenzies... as other people are."[23]

The press release "A Century Awash in Blood" states that while past "centuries have seen deadly feuds..., this century, scholars say, has been singularly conducive to mass murder [less in the name of fascism than Marxian science — estimated at over 85 million]".[24] The further report that "life lost value in... ideological blueprints to remake society" seems lost on many philosophers, as suggested in *The Oxford History of Western Philosophy.* Though admitting that some philosophers have

---

20 Cf. Laird Addis' "Memorial Minutes" on Gustav Bergmenn, *Proceedings of the American Philosophical Association* 61 (1987) 164, 165. Emphasis added.

21 Nicholas Rescher, "Where Wise Men Fear to Tread," *American Philosophical Quarterly* 59 (1990) 259.

22 Cf. Suppe, *The Structure of Scientific Theories,* p. 7.

23 Rescher, "Where Wise Men Fear to Tread," 59.

24 See A. Levinson, "A Century Awash in Blood — Spotlight: Genocide," *The Associated Press* 17 Sept. 1995, for both this and the next quote. Also, S. Courtois, Director at the National Center of Scientific Research in Paris and a self-avowed former "leftist", coauthored the research volume *Black Book of Communism* (NCSR, 1997) which specifies at least 85 million murders in the twentieth century in the name of Marxism: From the organized extermination of Soviet peasants to Stalin's purges and planned famine in the Ukraine, prior to World War II, to Soviet gulags and rape of Eastern Europe as well as China's Cultural Revolutions, Castro's repression, and Pol Pot's Killing Fields after World War II. Archives remain to be seen on such regimes as North Korea and North Vietnam.

recently had second thoughts about glibly dismissing medieval metaphysics, including proofs of God, this seminal history of philosophy notes:[25]

> In the 1970s, paradoxically, Marxism in the East was universally taught and almost universally disbelieved, while Marxism in the West was taught... but [typically] to an audience of passionate believers. Now, of course, as a result of the dissolution of the Soviet Empire... the institutional support for Marxist philosophy in Eastern Europe has almost totally collapsed. That philosophy must depend for its survival on... its devotees in the universities of the West.

The above invites reconsideration of a millennial reasoning with its apex in Thomas. Are modern anti-metaphysical philosophers better able to explain scientific truth, and its relation to ethics and politics, than by a voluntary Cause? If they are, is it too abrupt to ask why they did not come forward rather than often contribute scholarly minutia, as the academic community became politicized, or positively worsen the politics?

The word "minutia" is not meant to ignore contributions of modern philosophers. The point is that going about the business of teaching philosophy and doing esoteric research, as if there was no politicization of virtually every endeavor — signified by an ideological notion that even the personal is political, is a bit like "playing philosophical fiddle" while Rome burns. And this may be the best scenario. The worst, suggested by Kenny, Bergmenn, Rescher and others, may be more accurately depicted, dishearteningly, as the fire being positively fanned by either the ideological commitments or sheer opportunism of many philosophers themselves.

In sum, attention to metaphysics, for articulating a "truth" in physics, has been either rejected or ignored by the very philosophers who once held physics to be ideal knowledge. By an evident modal knowledge in science, a case is now made for scientific truth and for the Cause. Also, in terms of the Cause, a realist view of the "truth" is strengthened as well as shown to bear on ethics and politics. This important goal is related to Thomas' second way.

---

25 A. Kenny, "Afterword," *Oxford History of Western Philosophy*, Ed. A. Kenny (Oxford: Oxford University Press, 1994), p. 368.

Chapter 9

Experience to Modal Scientific Reasoning

In seeking to defend the soundness of Thomas' second way, several points need to be stressed. First, as previously noted, this way is selected among the other four by virtue of both embracing a notion of "efficient causality", beyond that of Aristotle, and finding limited expression in a causal principle of modern science. Second, the words "a defense" denote the indefinite article because there may be other defenses. The following involves elementary modal logic as it relates to causally dependent phenomena of traditional science and recent elements of chaos theory in physics.[1]

However, in addition to other well-publicized defenses which do not consider such a theory or logic, there may be other analyses that have been overlooked. One involves similarities between a Thomistic knowledge *simpliciter* of Nature's dependent nature and recent philosophies of science that suggest that to accept historically generated domains of theoretical truth, dependent on a uniformity of Nature, is to accept a broad concept of "cause" in a domain on which the others rest. Another analysis might even stress the dependence

---

[1] Scientific modalities discussed herein were the basis for my more expanded critique of modern epistemology in *From Physics to Politics* (NJ: Transaction Publishers, Rutgers -- The State University, 1999). Its last chapter was the core of my presentation, invited by physics professor Maria Falbo-Kenkel, to the Scientific Research Society of Sigma Xi at the Natural Sciences Center at NKU on October 8, 1997.

of physical things on a First Cause, reminiscent of the Greek "*aitia*" (a Cause that inspires), whose ultimate end is unknown.

Having acknowledged such analyses to which there will be further reference for clarifying and strengthening this defense, let the defense begin by noting that a central issue in Thomas' deductive argument is its soundness. Since soundness presupposes validity — that a conclusion follows with logical necessity, what is at issue is not this sort of necessity but rather the truth of the premises.

Shortly, it will be argued that the conditional premise cannot be rejected as false when a rudimentary modal reasoning of science is affirmed to be true. Support for the truth is indicated by eminent scientists as well as by philosophers such as Wittgenstein and W. H. Walsh. The latter illuminate a peculiarly strong nature of the modal truth. Before exploring the truth and scientific reasoning, the general relation of modal logic to Thomas' proof is examined.

The Proof and Modal Logic

Given a conversion of the modal impossibility into a necessity as previously noted, several points may be made about the last half of the second way after briefly clarifying its argument :

(i) Necessarily if there is no First Cause ($\sim F$), there are no second causes ($\sim S$).
<u>There are second causes (S).</u>
There is a First Cause (F).

(ii) Necessarily $\sim F \longrightarrow \sim S$
S
$\therefore$
F

The Proof's Modus-Tollens Form

First, it needs to be reiterated that the first premise may also be expressed "It is impossible for $\sim S$ to be false when $\sim F$ is true". Thomas holds that our experience of the very nature of second causes (S) is such that we are led ineluctably to hold that without a First Cause (F), the causes cannot exist. Modern logicians know that the syllogism per se has a valid form. The primary issue concerns the truth of the premises which, given the valid form, ensures the truth of the conclusion. Again, an apparent truth of the conditional first premise, that renders noncontroversial the second ("*S*"), is addressed shortly in terms of the "later" Wittgenstein's own insights.

Second, a modal fallacy is not committed in Thomas' *modus tollens* whereby from "Necessarily if ~*F* then ~*S*, and *S*", it is concluded that "*necessarily F*". Let "necessarily" be symbolized as follows:

$$\Box$$

The fallacy underscores that the conditional,

$$\Box(\sim F \rightarrow \sim S),$$

is not equivalent to either "If *necessarily* ~*F*, then ~*S*" or "If ~*F*, then *necessarily* ~*S*" which are represented respectively as:

$$\Box \sim F \rightarrow \sim S \quad \text{or} \quad \sim F \rightarrow \Box \sim S$$

For example, we have seen that "Necessarily if there is no God, there is no world" means that it is *impossible* for "there is no world" to be false when it is true that "there is no God". However, in "If necessarily there is no God, there is no world", it is *possible* that "there is no world" is true or false even though "necessarily there is no God" permits only true true-values. Though the antecedent of the latter conditional is "stronger" than that of Thomas' conditional, it is weaker than his conditional in the sense that his conditional does not allow for there being a world when there is no God.

Third, $\Box(\sim F \rightarrow \sim S)$ and $\Box(F \rightarrow S)$ are not equivalent. Given that "Necessarily if there is no First Cause, there are no second causes", we cannot validly infer that "Necessarily if there *is* a First Cause, there *are* second causes". This fact may not weaken Thomas' general theology because a First Cause *qua* God need no more have created the universe than a universe that is finite as opposed to infinite. Nevertheless, when

Thomas says that the union of a lover and beloved involves both the *presence* of the lover and *union* of affection (I-II, 28, 1), he indicates several striking things.

First, there is an implication for the sacraments. Thus for example, the Eucharist involves the *union* and *presence* of Christ with His beloved by a transformation of the bread and wine into His body and blood. Interestingly, this marks a "difference between the Catholic and... classical Protestant theology of the sacraments".[2] Second, the nature of God as Love is such that He must will reciprocated love. Accordingly, He would have had to create His beloved. Thomas suggests that the proper use of will, sensation, and intellect, in view of modal notions other than epistemic ones, as explained shortly, enable the beloved to grasp an impossibility that both the world and the beloved might not exist.

Logic and Revelation

Some points about a revealed God are summarized as a prelude to a First Cause. A brief comparison of the first and second ways may be helpful. The first underscores that experience and reason must be complemented by revelation for grasping that the universe was created, whether or not it existed coeternally with God, in contrast to Aristotle's pagan thought in which there was no creation.

Though the medieval scholar Frederick Copleston notes that Mohammedan believers were addressed and that "God is recognized" in the proof's conclusion,[3] we recall that a First Mover causes universal change and not existence. Thus this Mover is similar to an Aristotelian Unmoved Mover that exists as a unity with the things it moves.

The second way, by comparison, does not establish the existence of the revealed God. In going beyond this God merely causing universal change to causing the existence of the world, however, Thomas' argument is linked modally to a divine

---

[2] See Kreeft's insight in *Summa of the Summa*, p. 439, fn. 165.

[3] F. Copleston, *A History of Philosophy* 2: *Medieval Philosophy II* (NY: Image Books, 1962), p. 62.

Creator and not to anything which could be interpreted as an inseparable Aristotelian "form of the world's substance". At the same time, experience and reason need to be complemented by revelation for grasping that the world has not existed coeternally with God; that He created a finite world in which "time", following Augustine, is no less a creation than it.

Importantly, disbelievers as well as believers may reason from an experienced world to a First Cause without revelation. Believers in particular may also reason from God's cupidic nature to the existence of the world. Given the world's existence, this reasoning may seem trite. Yet it *(i)* explains perennial experiences such as uncondi-tional love and free will that elude other religions and philosophies, *(ii)* allays an urge, addressed by existentialism, to seek meaning in life from political ideologies, and *(iii)* diminishes a despair denoted by the word *"angst"*, in con-temporary continental thought, that has found its way into popular culture.

Given that God qua Love *"cannot* deny Himself", believers may grasp that their creation as the beloved had to obtain. The necessity "had to" and impossibility "cannot" pertain to a spiritual modal reasoning. On the one hand, Thomas' proof is a short step to the revealed God. A scholar who had disbelieved, but came to believe after accepting the proof, might grasp the impossibility of there being *no* world $(\sim S)$ when there is a First Cause (F). That is, "Necessarily if there is F, ... S." Accordingly, the definition

$$\Box(p \longrightarrow q) =_{\text{df}} (p \Rightarrow q)$$

is reintroduced for a formulation in which the conditional may be taken with "F" to yield a modal analogue of *modus ponens* with its inference to S:

$$(F \Rightarrow S) \wedge F / \therefore S$$

Since Thomas' argument is not a conditional with consequent as conclusion, the symbol "$/ \therefore$" may be better for denoting that a conclusion (S) follows validly from the premise $[(F \Rightarrow S) \wedge S]$:

In other words, to say that it is impossible for S to be false when

$$(F \Rightarrow S) \wedge S$$

is true is to say that the latter logically implies S. That is, S follows necessarily.[4] On the other hand, when S involves a "cupidic dependency", the argument for F could be of a First Mover or God reminiscent of Aristotle's *Metaphysics* (1072b): Necessarily if there is no loving God $(\sim G)$, there is no world whose motion is produced as being loved $(\sim M)$. And there is the latter (M), therefore the former (G).

These deontic modalities are significant. Perhaps, contrary to a traditional distinction of faith from reason, a Thomistic reasoning indicates that reason is more entangled with faith than imagined. Whatever is their relationship, the following discussion is primarily concerned with physical reasoning as it is related to alethic and epistemic modalities. Whereas the former are generally expressed by sentences with the words "necessary" or "impossible", the latter also involve words such as "knows" or "doubts". Intriguingly, Wittgenstein acknowledged a mode of doubt that bears on Thomas' premises. Consider his insight on when it is reasonable to doubt.

Epistemic Status of Premise's Truth

An important insight related to modalities is made in Wittgenstein's *On Certainty*. After noting that "only in such-and-such circum-stances does a reasonable person doubt...," he asserts:[5]

> The procedure in a court of law rests on the fact that circumstances give statements a certain probability. The statement that, for example, someone came into the world without parents wouldn't ever be taken into consideration.

What can be meant other than that there are no circumstances which could give the statement an improbability inducing *mere* doubt? Wittgenstein adds: "There are cases where doubt is

4 Cf. modal terminology in H. Kahane and P. Tidman, *Logic & Philosophy* (Belmont, CA: Wadsworth Publishing Co., 1995), p. 385.

5 L. Wittgenstein, *On Certainty* (NY: Harper & Row Publishers, 1972), p. 42e, #335, emphasis added. Being subject to interpretation, the word "parents" need not be countered prima facie by new biological developments.

unreasonable, but others where it seems logically impossible. And there seems to be no clear boundary between them."[6] This is a remarkable statement by a former icon of positivism — one who later inspired the dictum that whatever can be said can be said clearly. That he noticed an unclear boundary between unreasonable and logically impossible doubt, further underscores his uniqueness in a positivist Anglo-American tradition. The latter typically ignores any middle ground between empirical (*synthetic*) and logical (*analytic*) truth.

Further, the words "analytic" and "synthetic" are Kant's terminology. It is interesting to consider an understanding in the Kantian tradition of what may lie between empirical falsity and logical impossibility. In struggling with a Humean-influenced positivism in his *Metaphysics*, W. H. Walsh argued for the middle ground of a "categorial mistake."[7] There is the mistake in regard to a dropped quarter. A supposition that it disappeared and went clean out of existence lies between an empirically false "material mistake" of thinking it rolled left when in fact it rolled right and a logically impossible "formal mistake" that it rolled both right and left.

Clearly, there is a strong resemblance of "unclear boundaries" and categorical mistakes to modalities. A categorial modality might be "Necessarily if *X* ceases to exist, it does not become nothing" or "It is impossible for *X* to become nothing when it ceases to exist". The latter reflect general modalities that may lead to an inference of Thomas' conditional. They do not have the conditional's specificity of everything being ultimately dependent on a First Cause and of causes being both voluntary and natural. But categorial modalities find specific expression in scientific ones such as the impossibility of persons coming into existence when there are no prior reproductive processes. And a necessity of the processes renders *more* than unreasonable various empirical claims. The claims include the one that someone came into the world when there were no prior psycho-biological causes.

---

6 *Ibid.*, p. 59e, #454.

7 W. H. Walsh, *Metaphysics: An Exposition & Defense of A Controversial Branch of Philosophy* (NY: Harcourt, Brace & World, 1962), pp. 154-58.

These causes may have evolved from sheer biophysical processes, they from physical ones, and they from energy in terms of an astrophysical theory. But the causal series is still dependent in the sense of being insufficient to cause itself and begs modally for an uncaused cause that is not dependent. We can reason from the dependence to the conditional and the conditional presupposes the dependence. Yet the scientific modalities are not inferred directly since their discoveries are peculiar to different disciplines. Thus although the modalities presuppose a causal dependence, the conditional has a primary and ubiquitous applicability that they lack.

Some modalities may include our choices and thoughts such as those in the human sciences, and others may not such as those in the natural sciences. Yet how could they or the conditional not have a stronger epistemic status of "truth" than the truth-claims for which they set parameters?

The foregoing points may be summed up: *It is untenable to ascribe "truth" to empirical claims and not to modalities whose denials would, for Walsh and Wittgenstein, be more seriously mistaken than denials of either empirical truth or reasonable truth-claims. And though no particular circumstances tend to count against them, say in terms of a liberal verification principle,*[8] *they are not either meaningless or senseless in the sense of a speculative metaphysics without a perennial touchstone with our fundamental experience. The modalities have a stronger epistemological status than either empirical truth or truth deemed reasonable to believe.*

The Truth by Way of Science

Modal truth has been shown to have points of contact with a fundamental experience, to lie between empirical and logically necessary truth, and to set parameters for what is either more or less than reasonable to claim about experienced things. Our experience of their dependency finds expression in scientific modalities that are true in an unusually strong sense and that

[8] Simply stated, a liberal verification principle, as opposed to a conservative one, specifies that a sentence is meaningful if and only if something counts as evidence for or against its truth.

range from geology and biomedical research to astronomy to physics. While the modalities are akin to so-called "necessary laws of nature", they are noted as modal scientific truths for a continuity of language. And prima facie, as soon argued, they both strictly imply the truth of Thomas' conditional and render sound his argument.

Modal Truth in Geology

Geologists would naturally suppose that necessarily if certain metamorphic processes do not occur, marble is not produced. If geology students made a field report that marble was produced without the processes, they might evoke the valid response of their professors that "It's impossible marble was produced!" The response would indicate a modal reasoning reminiscent of Wittgenstein's "courtroom" example in which someone without parents, as the production of marble without the processes, "wouldn't ever be taken into consideration". Given the metamorphically dependent marble, the professors would not consider the possibility of its production when there are no processes. And their reasoning could be expressed "Necessarily if the metamorphic processes do not occur *(~O)*, marble is not produced *(~P)*":

$$\sim O \Rightarrow \sim P$$

Taken with the empirically true statement "Marble was produced"*(P)*, there is not deduced a modally fallacious conclusion "Necessarily the processes occurred". But there is yielded the true conclusion "They occurred" *(O)*:

$$(\sim O \Rightarrow \sim P) \wedge P \,/\therefore\; O$$

The patency of the conclusion should not obscure the fact that its truth follows validly if and only if both premises are true. The truth of the first premise is a primary modal scientific insight in terms of which Thomas' notion of "efficient causes" needs to be kept in mind. In discussing modern scientific modalities, it needs to be reiterated that second efficient causes do not create the matter *(ens)* of things but rather *(i)* depend on prior causes, *(ii)* cause change from one form to another, and *(iii)* sustain forms in a given series or species. It may seem peculiar that modern modalities have clearly denoted causes, *e.g.* metamorphic processes, in contrast to Aristotle's modalities such as "Necessarily if X is a man, it is an animal". Without denying a difference of the essentialism, to speak of "animal" and "man" as formal causes in *beings* is to speak of them as efficient causes that sustain "animal being predicated of man". A relation of efficient to formal causes accords with Augustine: "There is also an inner form... with an efficient causality deriving from... [an] intelligent nature".[9]

Having made these points, the marble's dependence brings to mind new bacterial strains whose morphological forms are modified. The modification may be considered by a teleologico-modal reasoning in medical research.

The Truth in Biomedical Research

When Dr. S. Levy at Tuft's University School of Medicine noted that any new antibacterial drug should be used judiciously because it is "*not possible* that resistance won't occur",[10] he gave modal and teleological voice to a reasoning in medical research: It is in the nature of "Bacteria [to] use a variety of evolutionary tricks to outwit antibiotics".[11] Even a perfunctory glance at medical news reveals a recognition by the medical-community of the *impossibility* of it being false that there will be sudden modifications of genetically unaltered bacteria into virulent strains (M) when it is true that new antibiotic drugs are used (D):

$$D \Rightarrow M$$

By material as well as formal analysis, the reasoning may afford the contraposit without appealing to a sentential-logic operation of Transposition which is disputed in modal logic. It is the nature of bacteria to survive by *moderate* adaptations to *gradually* evolving biophysical systems unless drugs are introduced. The introduction of penicillan was a classic cause of most affected bacterial strains becoming resistent within a half

---

9 St. Augustine, *City of God*, p. 266.

10 S. Levy, Tufts University School of Medicine, "Drugs Barely Keep Ahead of Bacteria," *Newsday-Ky Enq*, 3 Mar. 1995, No. 338.

11 *Ibid.*, No. 338.

decade. Accordingly, researchers might hold, all things equal, "Necessarily if there are no distinctively sudden modifications of bacteria into virulent strains *(~M)*, new antibiotic drugs were not used *(~D)*":

$$\sim M \Rightarrow \sim D$$

Similarly, biologists may hold that "Necessarily if a given organism exists *(E)*, it has parents *(P)*", as well as that "Necessarily if there are no parents *(~P)*, a given organism does not exist *(~E)*:

$$\sim P \Rightarrow \sim E$$

General Physics

The reasoning is equally applicable to physics and bears on astrophysics. When physicist B. Yavorsky says that the first law of thermodynamics "asserts it is *impossible* to construct... a periodically funct-ioning machine that would do more work than energy supplied to it from outside",[12] the impossibility may be recast as a necessity: "Necessarily if periodically functioning machines have no outside energy sources *(~O)*, their work does not exceed their internal energy *(~I)*":

$$\sim O \Rightarrow \sim I$$

Taken with "Their work does exceed the internal energy" *(I)*, there is again a Thomistic argument form of a *modus tollens* for the conclusion "They have outside energy sources" *(O)*.

Despite various interpretations of the laws of thermodynamics, the strong epistemic status of the truth of the laws is acknowledged by astrophysicist Victor Stenger. In reference to COBE-Satellite data that corroborates an inflationary Big-Bang theory,[13] he admits that he must show how the theory is compatible with the laws for holding that our "currently existing structure of the universe, including the laws of physics, could very well have been spontaneously generated after Planck time... [10^{-43} sec by the fluctuation of a black hole with

[12] B. M. Yavorsky and Yu. A. Seleznev, *Physics: A Refresher Course*, Tr. by G. Leib (Moscow: MIR Publishers, 1979), p. 165.

[13] See University of Hawaii physicist and astronomer V. Stenger's "The Face of Chaos," *Free Inquiry* 13 (1993) 13-14, emphasis added.

radius 10^{-33} of a centimeter, about 15 billion years ago]". Furthermore, the spontaneity may seem inconsistent with a Thomistic dependence. But our experience of Nature, as opposed to mere logical possibility, would lead us to infer that the fluctuation was dependent on the black hole whose existence is itself dependent, say on a previous collapsed universe or other phenomenon as, by analogy, currently existing holes are often held to be caused by collapses of massive stars: "Necessarily if there was no cause of the black hole $(\sim C)$, it would not have existed $(\sim E)$": $\sim C \Rightarrow \sim E$. Taken with "It existed (E)", there is yielded:

$$(\sim C \Rightarrow \sim E) \wedge E \mathbin{/} \therefore\ C$$

To affirm that the cosmic black hole was not causally dependent is to affirm that a phenomenon, from which arose the universe, is unlike the universe and everything else in it in a rudimentary modal way. However, Heisenberg's Uncertainty Principle and Einstein's relativity are used to explain how the fluctuating hole expanded to a protonic-size sphere with a "vacuum energy equivalent to all... presently visible mass".[14] Thus to deny that the modal necessity $\sim C \Rightarrow \sim E$ applies, in virtue of a logical possibility the hole has no cause (E when $\sim C$), is to deny prima facie the applicability of a relativistic equation between dependent phenomena of mass and energy that renders intelligible the explanation.

Science and Philosophic Confusion

In expressing a modally necessary causal dependence of all phenomena (S) subject to scientific inquiry, Thomas' conditional $\sim F \Rightarrow \sim S$ does not hold that without a given first phenomenon there are no second ones. It does not suggest, by analogy, that without the first parents of Adam and Eve there could be no human race. Scientists cannot determine, by reason or experience per se, whether our race or any other phenomena consist of an infinite or a finite series. The conditional admits of either series as a logical possibility. But given our *experience* of the

14 *Ibid*, pp. 13-14.

world, there is an ontologico-modal dependence on what is not dependent that is not obviated by the possibility. A possible epistemological regress and the impossible ontological one, to put it another way, should not be confused.

Perhaps, confusion has been exacerbated by an ambiguity of the word "contingent". Since the time of Hume and Kant, philosophers have often begun with a contingency in which there is a *logical possibility* that anything might not exist. This idea of contingency may have obscured an experienced one. Some of the philosophers, who may include Wittgenstein and Heidegger, may have reasoned from experienced things to a *modal possibility* of their "not being" in virtue, precisely, of their not being "necessary" because they are dependent. The dependence per se, of course, was characteristic of Thomas' first three of proofs of God. That is, an inferred modal possibility was evidently confused, by interpreting it through epistemic lenses of the paradigm, with a logical possibility that was both devoid of the experience and not inferred.

This confusion may bear on Thomas' third proof of a Necessary Being by way of possible beings. There may be a misunderstanding that the proof begins with either the logical or modal possibility of beings "not being". These alternatives would exclude a necessary Being in the sense that a logically necessary one cannot be inferred from a logically possible one and, from a modally possible one, there can be no inference to one who in fact exists. But Thomas does not reason from modally or logically possible beings. He begins with the experience of their coming to be and perishing for inferring they are dependent "possible beings". Arguably, in being based on experience, his inferences yield an impossibility there are possible beings *(P)* when there is no Necessary Being *(N)*: $\sim N \Rightarrow \sim P$.

If "possible being" is understood in the Humean-Kantian tradition as a logical possibility that whatever exists might not exist, an inference may not merely be confused with its experiential origin. A False-Dichotomy Fallacy may be committed by ignoring a modal necessity that is weaker, epistemologically, than logical necessity but stronger than empirical contingency. These errors would lead to an *underestimation* of Thomas' proof as valid, but not sound, and to an *overestimation* of radical possibilities. The latter stem from stressing empirical truth, as truth alone about reality, influenced by empiricism. We recall that Hume, in taking his empiricism seriously for a moment, checked frantically to see if the world was still there! Truth based on brute contingent possibilities, not on any modal impossibilities about the world, was alone admitted. In view of modal scientific reasoning, how can the odd behavior be tied to an irrational denial of both the truth of Thomas' conditional and soundness of his argument?

Soundness of Argument

To deny "$\sim S$ cannot be false when $\sim F$ is true," which expresses Thomas' conditional $\sim F \Rightarrow \sim S$, is to deny that $\sim F \Rightarrow \sim S$ cannot itself be false when S is true: $S \Rightarrow (\sim F \Rightarrow \sim S)$. For, bearing on the issue of soundness, $S \Rightarrow (\sim F \Rightarrow \sim S)$ and $\sim F \Rightarrow \sim S$ are equivalent. Precisely, the modal sentence "$\sim F \Rightarrow \sim S$ cannot be false when S is true" expresses that since second causes cannot cause their own existence, they cannot exist when $\sim F$.

Since S may denote either a dependent causal series presupposed by science or scientific modalities, as soon noted, affirming S and thus $\sim F \Rightarrow \sim S$ is *more* than reasonable. But taking $\sim(\sim F \Rightarrow \sim S)$ with $S \Rightarrow (\sim F \Rightarrow \sim S)$, to infer $\sim S$, is not merely unreasonable. It conflicts with science itself. While truth tables may be strictly inapplicable to modalities, they may afford a more intuitive grasp of the equivalence ($\Leftrightarrow$):

$(\sim F$	$\Rightarrow$	$\sim S)$	$\Leftrightarrow$	$[S$	$\Rightarrow$	$(\sim F \Rightarrow \sim S)]$
F t	**T**	F t	**T**	t	**T**	T
F t	**T**	T f	**T**	f	**T**	T
T f	**F**	F t	**T**	t	**F**	F
T f	**T**	T f	**T**	f	**T**	T

One may object that whereas the equivalence may hold in modal systems as strong or stronger than S4, it does not hold for weaker alethic ones (perhaps T) and does not obtain for subjunctives and nomic laws of nature. But we need not delve into the systems to note that the conditional is not one of the laws. Though the laws and conditional presuppose the causal dependence, as previously discussed, we can reason from the dependence to the conditional but not to a given law per se in the same immediate way.[15] Thus the ubiquitous dependence includes our thought. In being applicable to it in a way reminiscent of the Laws of Thought without triviality, the dependence and conditional apply prima facie by strict implication to thought about the systems as well as to the systems themselves. While our consciousness of thought and freedom to think involve a voluntary causality without a natural cause per se, the latter cause is a necessary condition for thought as well as for its manifestations and they are part of a second-cause series.

Further, among different interpretations of them, including that the systems are different and not weak or strong per se, I suggest one where they may overlap as do systems of classical and quantum logic for macroscopic and microscopic domains. Hence, even by this analogy, the conditional may apply to different modal systems as may a formula in different systems of physics (*e.g.* as may $v_i = P_i/m_i$).[16]

These points may be made with a hope that modal systems may someday be more clearly explicated without relinquishing our apparent case, as by analogy "mathematical proofs were... rationally acceptable long before logicians could give a rigorous account of them".[17] Prima facie, to say scientific inquiry presupposes the impossibility there are second causes sufficient to cause themselves $(\sim S)$ when there is no first cause that caused them $(\sim F)$ is not only to say $\sim F \Rightarrow \sim S$ but: Necessarily if there are second causes $(S \Rightarrow)$, this fact is impossible $(\sim S)$ when there is no first cause $(\sim F)$: $S \Rightarrow (\sim F \Rightarrow \sim S)$. And $\sim F \Rightarrow \sim S$, as noted in Appendix I, may be articulated apart from a given modal system such as S4.

Moreover, since S finds expression in modal scientific

---

15 I am beholden to Prof. Del Ratzsch for the above possible objection. See Appendix *I* for a Kripkean interpretation that may render irrelevant whether Thomas' conditional and various laws are in S4..

16 Cf. Haack, *Philosophy of Logics* (London: Cambridge University Press, 1978), p. 223. While a comparison of modal systems to physics is noted, the overlapping domains and law are not.

17 Peter Geach, *God and Soul* (NY: Schocken Books, 1969), p. 85.

necessities, "S_m" may designate this fact without skewing its meaning: $\sim F \Rightarrow \sim S_m$ cannot be false when S_m true. Since a denial of S_m lies in epistemic strength between logical falsity and either empirical falsity or unreasonableness, the denial would not seem merely unreasonable but irrational. It follows that, to avoid an evident irrationality is to affirm the truth of S_m and therefore of $\sim F \Rightarrow \sim S_m$.

Finally, another way to view the matter is to distinguish strict ("$\Rightarrow$") from material implication ("—>"). A modal sentence "'Art is an artist' (A) strictly or logically implies 'Bill is big' (B)" reads "$A \Rightarrow B$". Though $A \longrightarrow B$ is true if A and B are true, for instance, $A \Rightarrow B$ is false despite the truth-values of A and B since "A / ∴ B" is not a valid argument even if A and B *happen* to be true.[18] Prima facie, however, $\sim F$ / ∴ $\sim S_m$ as well as S_m / ∴ $\sim F \Rightarrow \sim S_m$ are not only valid but sound. This is to say that, given that $\sim F \Rightarrow \sim S_m$ cannot be false when S_m is true and S_m is a modally necessary truth, the truth of the conclusion $\sim F \Rightarrow \sim S_m$ follows necessarily. When taken with S_m, there is the sound argument: $S_m \wedge (\sim F \Rightarrow \sim S_m)$ / ∴ F.

In sum, modal scientific laws bear on Thomas' reasoning to God, understood as a First Cause, wherein it is impossible that there is a world of second causes when there is no First Cause. This reasoning may be expressed in terms of a scientific proof of God in the sense, precisely, that well established scientific laws (S_m) strictly imply that the laws cannot obtain ($\sim S_m$) when there is no God ($\sim F$). And in recasting the impossibility as a necessity, it may be stated: Necessarily to accept modal scientific laws (S_m), expressing relations of second causes such as a law of thermodynamics, is to accept that necessarily if there is no First cause ($\sim F$), these laws do not obtain ($\sim S_m$). Since the laws do obtain (S_m), there is a First Cause called "God" (F).

---

18 See Kahane and Tidman, *Logic & Philosophy*, p. 385, for this comparison of material and strict implication.

Chapter 10

Reasoning in General Physics and Metaphysics

The proof of a First Cause, in terms of modal scientific laws (S_m) leads to other considerations. They include how the Cause is related to a teleology of recent bio-physical and physical theories. Before exploring how the theories may embrace a *telos* of self-organizing purposive phenomena that reflect a traditional metaphysics, it needs to be noted that theories of physics presuppose a metaphysical theory of truth (epistemology). For example, where "*p*" is any truth-claim, to say "*p* is true" or "*p* is known" is to be able to say what is "knowledge" or "truth". Thus alhough Thomas embraces a realism wherein statements are true by virtue of corresponding to reality, a tenability of the realism might seem to rest on knowing a possibly infinite number of modal truths.

That is, to affirm the truth of scientific necessities (S_m) is to affirm both the truth of Thomas' conditional and the soundness of his argument. But there might be a supposed dilemma of knowing that a possibly infinite number of the necessities (S_ms) are true. In terms of sentential logic, the falsification of one conjunct falsifies any conjunction. Thomas' reference to "no case known", already examined, invites the response of an atemporal knowledge *simpliciter* of interrelated necessities or impossibilities.

Understood traditionally as knowing "many by one"— say many phenomena of light in terms of one theoretical entity "wave" (λ), a notion of "knowledge *simpliciter*" in the current philosophy of science may serve to strengthen its tenability.

For example, Cornell physicists F. Rohrlich and L. Hardin suggest a knowledge *simpliciter* with respect to an approximate truth of theories of physics in given domains of a historically-generated sequence of interrelated "nested domains".[1] An approximate truth about them is typically so well tested that the theories could only turn out to be entirely false on the *more* than unreasonable supposition that Nature itself might suddenly and radically change. Is to accept the nested domains of truth to accept the truth of a modal necessity which may be common to them, say in terms of a general cause of a "nest" that nests the others?

Thomas' notion of "efficient causality" is applicable prima facie to a "nest" of nested domains by virtue of its inclusion of voluntary as well as natural causes of modern causality. Thomas ties a First Cause *qua* God to our knowledge of natural objects: As "the natural objects of knowledge are prior to our knowledge..., so the knowledge of God is prior to natural things, and is the measure of them" (I, 14, 9).[2] The medieval scholar Peter Kreeft aptly refers to this passage for noting that the medieval world view includes, surrounds, and expands on a "positivistic-materialistic-empiricistic-scientific-secular world view... [of] modern man".[3]

In causing our existence as rational efficient causes who obtain knowledge of which we are not the the mere "measure", God as the First Cause renders intelligibly objective scientific knowledge. At the same time, in reducing reality to natural phenomena, modern defenders of an exhaustive causal determinism do not merely disallow for any objective "measure" by excluding an immaterial self-conscious reflexive intelligence.

---

1 F. Rohrlich and L. Hardin, "Established Theories," *Philosophy of Science* 50 (1983) 603-17.

2 Though Thomas is associated with Aristotle, this notion of "God" brings to mind Plato's *Cratylus* in which true ideas are related to ideas of God. Since human ideas can be *incorrect*, there must be "a power more than human [that] gives things their first names... [and] a standard which shows the truth of things" (438c), where "truth" is not ascribed to *things* but rather to *ideas* about them which begs for a notion of the mind of God (407b).

3 Kreeft, *Summa of the Summa*, pp. 139-140, fn. 118.

They surreptitiously posit natural causes of their own truth-claims and counter-claims with no rational way to assess which are true because even the assessments would be determined.

And importantly, modal knowledge of first and second causes affords an inference, by the inference rule of Simplification, to a modern causal principle. The inference is needed for making sense of a scientific truth that presupposes the truth of the principle, whether for the methodology of either a limited or thoroughgoing determinism. What Kant held to be a *metaphysical* principle, which was a priori and synthetic without known truth, had a *metaphysico-modal* truth. The latter had more epistemic strength than empirical truth.

A modal truth that all things have voluntary and natural causes affords an inference to the modern principle. But Kant's view does not permit an inference to voluntary first-causes with which the principle is deemed inconsistent. The causes include persons as psycho-biophysical beings who freely initiate causes and who may grasp that there are historically nested scientific domains of which there is knowledge.

There is difficulty in seeing how a knowledge simpliciter of efficient causality, which underlies the domains, could be seriously challenged apart from the specter of Nature itself either ceasing to exist or radically changing. An instance of considering such a radical change would be exemplified by astrophysicists who express astonishment over the existence per se of a neutron star[4] — as if an "attribute" of its possible nonexistence was as natural to suppose as the existence stemming from its causally dependent nature, namely on a supernova explosion of a massive star.

Even if future theories modify our understanding of stars, it would be most peculiar to suppose that the dependent nature of phenomena would be abandoned. For there would be abandoned a notion that renders intelligible scientific inquiry itself. Events might simply be understood to occur for no reason as well as from "nowhere" and out of "nothing". These considerations warrant further exploration of the metaphysics which supports a Thomistic reasoning.

---

4 Recall Heidegger, *An Introduction to Metaphysics*, p. 25.

The Argument in Astrophysics

In considering teleological knowledge, in view of Thomas' argument, another element in his epistemology needs to be addressed. This element is related to a teleology of recent astrophysics. It is beyond present purposes to expand adequately on the element. Suffice it to say that the element involves a metaphysics of the universal (*essentia* or form) "in" individual phenomena which permits an Aristotelian *epagoge*.

There is no formidable problem prima facie in holding that perception together with memory and intellect enable scientists to inductively "abstract" universals in an "intellection process"; to grasp not only essential natures *qua* forms that, via historico-theoretical developments, yield formalized theories but their dependent nature. That is, in the most primordial sense, phenomena are essentially efficient causes in terms of which substantive modal conditionals are formulated.

The conditionals, in different sciences, have in common a causally dependent nature of phenomena. Thomas' conditional is a related conditional *writ large*. With no composition fallacy as already noted, the conditional expresses an impossibility of Nature itself existing when there is no Efficient Cause, that is not itself dependent, which causes it.

This point is not a broadside defense of a methodology which precribes intuited first-principles (*archai*). The latter were superseded by subjecting, what are often, complex counter-intuitive theories to empirical tests that may count against them. Rather, the point underscores that theories are coordinated with both perception and a broad common-sense (*sensus-communis*) experience. The experience may fruitfully issue in modal claims resting on metaphysical-teleological purposes of phenomena. The phenomena, acknowledged or not, include self-organizing biophysical systems.

Given that a modal reasoning from physics to biology often reflects an understanding of observational things in both formalized and unformalized theories, one might gather that the applied logic of "contingency", whatever *is* might *not be*, compares unfavorably and seems unnatural. And we need bear in mind that modern philosophies, from pragmatism and positivism to linguistic analysis and phenomenology, have been aggressively exploited, often naively and without due epistemic caution, for criticizing Thomistic metaphysics in general and a teleological causality in particular.

Keeping in mind that the causality involves an autonomous rational order apart from deterministic laws, there may be elaboration on Stenger's reference to COBE-Satellite data that corroborates an inflationary Big-Bang theory. This theory permits the universe's protonic-size expansion in 10^{-42} sec, with a vacuum energy equivalent to all present mass by Einstein's general relativity, from a quantum fluctuation in accord with Heisenberg's Uncertainty Principle. But the fluctuation of a black hole with maximum entropy or peak disorder implies an expansion from disorder to order by "natural processes of [autonomous] self-organization and even... a kind of Darwinian natural selection among... possibilities".[5]

These possibilities are linked teleologically to formations of sub-cosmic systems and their subsystems. Each system is caused by previously evolving systems and has an equilibrium dependent on its self-organizing adaptation to parameters of possibilities caused by other systems. For example, planetary systems adapt to both their outer evolving solar systems and iner physical, and possibly biophysical, subsystems in terms of a self-organizing purpose of equilibrium.

Biophysical and Political Systems

Such purposive adaptability, much more choices endemic to human biological systems, have found general expression in the family of quadratic maps:

$$Fu\ (x) = ux\ (1 - x)$$

Programming a computer to iterate the function $f(x) = 4x(1\text{-}x)$ with input of random numbers between 0 and 1 yield dramatically different behavior. Values unpredictably repeat and do not repeat as well as "wander aimlessly about the unit interval with no discernible pattern".[6] Though noniterated systems also

---

5 Stenger, "The Face of Chaos," p. 14.

6 R. L. Devaney, *An Introduction to Chaotic Dymanical Systems* (NY: Addison-Wesley Publishing Co., 1988), p. 3.

produce unpredictable results, researcher Pierre Berge prescribes a simplification of systems in which not only the equation $f(x) = 4ux(1\text{-}x)$ describes a period-doubling cascade "but... any nonlinear function of x".[7]

Thus in congruency with an astrophysical notion of a "purposive self-organizing" universe, planetary systems within parameters of subplanetary and solar systems may themselves have teleological adaptive bio-physical systems. They may find expression in a difference equation $Pn + 1 = kPn\ (1\text{-}Pn)$ for populations, where Pn = population after n generations, k is a constant, and a limiting value (L) of overcrowding and limited food is 1 ($L = 1$). Recast as a function in which $x = P_o$ (initial population) and $f(x) = kx(1\text{-}x)$, there results $P_1 = f(x)$, $P_2 = f(f(x))$, $P_3 = f(f(f(x)))$ and so on.[8] While a population's fate is subject to interacting stimuli, a given constant k involves the unpredictable behavior $kx(1\text{-}x)$.

Such behavior reflects an important understanding of causes in the social-political domain as well as in such domains as biophysics and astrophysics. Keeping in mind that the causality includes causes in terms of autonomous human choices — since they are voluntary "first" causes among other second ones, physicist A. M. Saperstein's "War and Chaos" supplements a linear set of differential equations for predicting instability and war with a mathematics that gives expression to human choices in light of various fears and needs of nations. The complexity results in nonlinearized equations, for interacting nations, which yield their long-term unpredictability. Invoking "chaos...," he notes, "is a valuable adjunct to conventional political science [for reflecting variables of the 'real-life world']".[9]

---

7 See Pierre Berge's remark in the seminal work *Order Within Chaos* (NY: John Wiley & Sons, 1984), p. 287, with Y. Pomeau and C. Vidal.

8 Devaney, *An Introduction to Chaotic Dynamical Systems*, p. 8.

9 A. M. Saperstein, "War and Chaos," *American Scientist* 83 (Nov-Dec 1995) 548-57. He speaks only of "need" or "fear" per se and allows for "deterministic rules". But their applicability to interacting notions of "autonomously acting people" shows that the rules are consistent with both choice and the long-term unpredictabilities in question.

What emerges with a spectrum of "worlds", from social-political and biophysical to physical and astrophysical worlds, is a continuity of independent analyses that variously posit a *telos* of choosing as well as self-ordering and self-organizing phenomena. These phenomena clearly reflect an inherent intelligence and purpose. Importantly, the interacting worlds of physical and biological systems are not merely interpreted in terms of stochastic assumptions by which there is either a purposeless randomness or probabilism in, say, the traditional manner of quantum mechanics.

In addition to being challenged by recent research in quantum cosmology, as illustrated by the work of Oxford physicist David Deutsch,[10] any probabilistic domain devoid of a *telos* is also implicitly questioned by physicists such as R. V. Jensen. He affirms interfaces where "nonlinearities are large and perturbation theory fails, classical theory exhibits chaos, and quantum mechanics is complicated..."[11] Thus for example, Jensen indicates how uncomplicated theories may be properly construed to ignore teleological complexities for limited purposes of predicting phenomena. But the construal does not obtain for understanding their phenomenal nature — where this difference makes a difference in appreciating the nature and limits of conventional theories.

This point does not assume either that conventional theories will be superseded by chaos physics or that the present discussion of this physics, herein, has even remotely touched on a host of other important considerations. Rather the point indicates an increasing significance of some issues for uncovering limitations of conventional theories, relationships between them, and an equally limited modern metaphysics. Another misguided objection to Thomas' metaphysics regards his "anthropocentrism".

---

10 See physicist D. Deutsch and philosopher of science M. Lockwood, "The Quantum Physics of Time Travel" in the *American Scientist* 270 (1994) 68-74.

11 R.V. Jensen, "Sigma Xi National Lectures 1994-1995," *American Scientist* 81 (Nov-Dec 1993) 603.

Anthropocentric Science

A Thomistic connection between the human sciences and physics is strengthened, despite objections to anthropocentrism. The latter is often seen as naively imposing human traits, from purpose to feeling, on nonhuman things. Exemplified in the most puerile way in terms of "dark cloudy skies" being depressed, anthropocentrism is typically viewed as being properly superseded by an impersonal physics devoid of medieval metaphysics. However, there is nothing childish in noting that if physical systems are construed as self-organizations of teleological adaptabilities to various possibilities, then scientists are warranted *a fortiori* in viewing social-political and biological systems in the same or similar way. For physical systems are their ontologico-evolutionary origin and physics their methodological model.

A Deference to the model is not hard to find in biosocial models with a teleological purpose and choice of persons. For example, it is stated by biosocial researchers that a "person is not a... passive recipient of social forces" but part of dynamic-impact models whose analyses in "physics can be found in... 'Statistical Mechanics of Social Impact'."[12]

Dynamic-impact models, no more than a primordial black hole's quantum fluctuation from peak disorder to an orderly formation of the universe, is related to efficient causality and not to a modern mechanistic determinism of exact or even inexact measurement. The Aristotelian-Thomistic paradigm for such models and order, which does not diminish its complex applicability, was the sculptor of marble or maker of a house. In terms of second causes, the latter broadly involved purpose, self-organization, and adaptability. Though a mathematized intricateness of recent teleological notions was not anticipated by Thomas, this fact does not obviate a revitalized significance of his concepts any more than novel interpretations of art, unforeseen by the artist, preclude their significance.

Today, there is a tendency to forget that the word "tech-

---

12 See D. MacPhee's reference to "Statistical Mechanics of Social Impact" in the *Physical Review A* (1981) 45, in his article "Directed Evolution Reconsidered," *American Scientist* 81 (1993) 554.

nology" has roots in the Greek concept "*techne*" in terms of which the arts and sciences were entangled. They were a *continuum* of painting, poetry, and sculpture at one end and geometry, physics, and astronomy at the other. The *kosmos* they depicted was no less understood as crafted in virtue of a principle of causal efficiency than the artistic-scientific works that were shaped by craftpersons —persons being understood as "voluntary second causes."

Persons as First Causes

In this sense, Thomas' notion that second causes cannot issue in an infinite regress without a Creator reflects an inference from our first-hand experience of ourselves as productive intellects with purposes. And there can be no intelligible talk of purposes apart from choices. In being origins of our activities, our choices comprise our ability to initiate change in matter. We change matter in the senses of both initiating change in its constituents and imposing new forms on it, and we experience ourselves in these ways as creators or first efficient causes.

Further, our incontrovertible experience of ourselves as such causes reflects an Aristotelian method of science. Scientific investigation should begin with the more familiar (*ourselves*) and proceed to less familiar organisms or entities, as variously noted in Aristotle's *Physica* 184a15 sqq, *Historia Animalium* 588a 20-5, and *De Partibus Animalium* 641b 10-15.

The objection that this methodology imposes anthropocentric idea on the world is not only challenged by chaos theory but by a scientific-philosophic environmentalism. In *Environmental Philosophy,* Valery Plumwood traces environmental destruction to Nature being construed as "bereft of qualities appropriate to the human side".[13] Nature is not "passive and lacking in agency and teleology... So what is called for... [are] alternatives to mechanistic ways of viewing the world".[14]

---

13 Val Plumwood, "Nature, Self, and Gender," *Environmental Philosophy*, Ed. M. Zimmerman, J. Callicott, G. Sessions, K. Warren, and J. Clark (Englewood Cliffs, NJ: Prentice Hall, 1993), p. 299.

14 *Ibid.* Yet, see criticism of Plumwood in Trundle's *Ancient Greek Philosophy* (278). A self-avowed feminist, she fails to note how many academic feminists have gone full circle from viewing Aristotle's thought as a paradigm "sexist rationalism" to insights reminiscent of his sciences.

There is, here, a stark contrast between viewing the world mechanistically and naturally whereby, in the latter, both Nature and human nature are bound together morally as well as scientifically. The human and social sciences provide for more than mere descriptions of our behavioral agencies and nature. Descriptions of what *is* the case about human nature are the basis for prescriptions of how it *ought* to be fulfilled as well as for teleological insights into Nature itself.

Consequences of Denying Such Causes

Having become a banal truism, modern objections to anthropocentrism also beg for a reiteration of evident incoherencies of a mechanically determined universe composed exhaustively of natural phenomena. Despite an inclusion of energy as well as the particles, reducing the cosmos to the phenomena results in a dogmatic philosophical materialism. The latter contrasts to a methodological materialism for limited purposes of inquiry. And it is tied to an equally misguided determinism. Consider this determinism after the materialism.

Immaterial realities such as freedom and thought may be disregarded by conventional physics, or by any of the social and natural sciences, for limited purposes of predicting and manipulating phenomena. But if disregard leads to denial, a coherent concept of scientific "truth" collapses. "Truth" is not ascribed to phenomena but rather to thoughts or statements about them. Therefore, if statements or thoughts are reduced to phenomena, there would only be phenomena and no truth.

A thoroughgoing determinism fares no better. Surely, Newtonian-Einsteinian equations are typically seen as deterministic of exactly measurable events and quantum physics of inexactly measurable ones. The universe may be likened to a cosmic machine with, respectively, precisely and imprecisely interacting parts. Yet if everything is exhaustively determined in terms of a transposition of medieval efficient causes into an invariable succession of mechanistic events, then even claims that theories are true would be causally determined. There would be no *freedom from* a deterministic spatio-temporal realm to rationally assess which claims were in fact true. Even the idea of factual "truth" would itself be caused with no way of assessing whether a caused denial of this "truth" was true.

Medieval Causality and Causal Truth?

This is not to say that a "working" assumption of a deterministic realm is wholly inapplicable to spatio-temporal domains modest in comparison to vast ones addressed by some modes of chaos physics. Though physicists acknowledge chaos in more classical domains, the assumption may be viewed as reflecting a truth about the domains by virtue of the irrelevance *ceteris paribus* of unpredictable self-organizations to short-term cosmic considerations.

As theories of physics disregard irrelevant properties — as classical mechanics does electric charges, the causal principle may also ignore the irrelevant; though as noted, a denial of certain human teleological traits renders problematic "truth". This returns us to a modern dilemma: "if skepticism is to be avoided, the exploitation of... 'causal' regularities in obtaining a posteriori knowledge must not require prior knowledge of those regularities".[15]

The skepticism reflects a falsely-dichotomized positivist idea, expressed by a verification principle, that there is only either empirical (*synthetic*) or logically necessary (*analytic*) truth that may be ascribed to a causal principle.[16] Typically,

---

15 Suppe, *The Structure of Scientific Theories*, p. 722. This dilemma is tied to the K-K Thesis: For "were this required it would be impossible to obtain our 'first' pieces of knowledge about the physical world, hence resulting in a pervasive skepticism. Denial of the K-K thesis thus is essential... enabling one to evidentially exploit the 'causal' regualrities involved in... perception in obtaining knowledge without knowing what those regularities are."

16 A conservative verification principle, that appeals to *synthetic* truth, reveals a positivist indebtedness to Kant. With the caveat that he speaks of "judgments" and not "sentences," *a posteriori* sentences might seem more applicable to empirical truth since they follow from sense experience. That is, *synthetic* sentences might be thought to merely synthesize different concepts, *e.g.* "events" and "causes" for a modern metaphysico-causal principle. But all *a posteriori* sentences are *synthetic* and some *synthetic* ones, though not *a posteriori*, may be potentially verifiable. The hypothesis for a forty light-year distant star "Star 51 Pegasus has an orbiting planet," made prior to 1993, synthesizes "Star 51 Pegasus" and "orbiting planet". While the hypothesis did not follow from experience, it was later verified by virtue of new technological developments.

the principle is understood as "all events have causes" or sometimes, somewhat more precisely, as "all events have exactly or inexactly measurable causes".

What is often ignored is how the dichotomy bears not only on skepticism over the causal principle but on scientific theories that presuppose it. On the one hand, given Hume's limitations of induction and its acceptance by Kant, empirical truth cannot be ascribed to the theories. And a notion that their truth is known analytically would make them trivially true and defeat the purpose of testing them by predictions.[17] On the other hand, by a material implication $(p \longrightarrow q)$, where q does not imply p, implied theoretico-empirical predictions that obtain do not imply empirically true theories.

Nevertheless and oddly enough, skepticism over the causal principle may find an answer in a problem of truth ascriptions to theories. Without reference to an alternative modal truth—though appeal to it would be a natural outcome, one may argue that as it would be *more* than reasonable to ascribe empirical truth to theories in virtue of their systematic predictive success, since the success would be inexplicable unless they truly reflect what reality is approximately like in given domains, it would be *equally* reasonable to say that the principle is limitedly true. For its truth is a necessary condition for the intelligibility of the accepted theoretical truth.

"Truth" in a Humean-Kantian Tradition

How could we coherently *affirm* that theories approximately describe properties, processes and relationships of phenomena and *deny* a continuity of past and future to which such time-bound relationships and processes are conceptually connected? To accept the processes and relationships is to accept prima facie a continuity of past and future in terms of future regularities expressed by the causal principle. Since it seems *more* than unreasonable to reframe from ascribing "truth" to the principle when it is ascribed to theories, why can we not specify "Necessarily if theories are limitedly true, the causal

---

[77] Such testing generally pertains to theories that can be fruitfully formalized.

principle is limitedly true"? Predictably, philosophers in the Humean-Kantian tradition may argue that a future falsification of theories is logically possible.

However, we need not delve into responses of philosophers who, as Imre Lakatos, appeal to modifying theories by auxiliary hypotheses in order to suggest another response. The very *behavior* of the philosophers in this tradition belies a disingenuousness of the possibility. For the possibility is posed in the buildings and technological environment, at conventions of philosophers and elsewhere, whose theoretical integrity is unquestioned. Presumably, the unquestioned theoretical integrity is tied epistemologically to theoretical truth and it to an unchanging nature of phenomena which is itself rooted in the philosophers' own incontrovertible experience. How can such experience bear on theoretical truth but not on the truth, or limited truth, of the causal principle?

The principle may be insufficient for a coherence of scientific truth. But it does not beget incoherence when it is understood as a limited expression of a Thomistic causality in which we are freely-choosing (voluntary) efficient causes among dependent second causes. Suppose the idea of such causes and modern causality are conceptually unrelated. Then, there could not have been, in the first place, any coherent transposition of efficient causes, as explanatory agencies, into invariably succeeding events in the Humean-Kantian tradition.

A point about this tradition may be considered in terms of Hume's challenge to an inductive assumption that the future will behave as the past. In addition to his actual assumption of a continuity of past and future for anticipating a future to which his skepticism is applicable, Stephen Korner has made a cogent insight. An induction of future behavior may be based precisely on our "expectation" without a regress of question-begging justified justifications. One need not argue that the future will behave as the past since it has in the past.[18]

One may add that our expectation, that the future will behave as the past, does itself rest on an experienced dependent

---

18 Stephen Korner, *Experience and Theory: An Essay in the Philosophy of Science* (London: Routledge & Kegan Paul, 1969), p. 194.

nature of phenomena. A *more* than unreasonable challenge to the experience and expectation would proceed with a *more* than unreasonable challenge to scientific truth. And the truth would include that of rudimentary modal scientific principles.

Another way of relating Korner's thesis of expectation with ordinary experience is by considering philosophical and everyday skepticism. In regard to the latter, Hume expresses no actual doubt. When he writes his skeptical thoughts, he must assume that his own reasoning as well as paper and table on which he writes will perdure in future moments in a way continuous with present and past ones. The continuity would only be actually doubted when not doubted philosophically, say after a vivid dream, and philosophically when not actually. Distinctions of genuine from disingenuous doubt turn against Hume his own distinction of himself as a doubting philosopher from an undoubting practical man.

A first step towards collapsing this influential distinction, for making sense of modal scientific principles and a principle of causality, begins by reexamining classical-medieval reasoning. Ironically, in the name of a Humean "critical thinking" used for logic courses and textbook titles, the reasoning has been thoughtlessly relegated to an archaic metaphysical methodology. An Aristotelianism of beginning with ourselves for understanding the world, that influenced Thomas, cannot be glibly dismissed as a pre-Copernican anthropocentrism beyond which we have "truly forged ahead".

Persons and Personal Creator

Thomas thought reflects an Aristotelian human agency more than Aristotle. Aristotle held in his *Ethics* 1139a that "The origin of human action — its efficient, not its final cause — is choice". The Unmoved Mover was not only an efficient cause inspiring a striving (*entelechy*) toward natural ends but, as a formal cause, an *impersonal* form of the world's substance.

Thomas admits of a logical possibility that the world was created by God with infinite time coeternal with Himself: "By faith alone we hold... that the world did not always exist" (I, 46, 2). But when he reasons that "by his free will man moves himself to act" and that a "first voluntary Agent" causes both voluntary and natural causes (I: 46, 2 and 83, 3), he suggests a stronger anthropic approach of an existential relation to a divine personal Cause. To accept ourselves as second causes is to accept the Cause, not of our voluntary action but of our dependent existence. We exist as natural bodies in which our free will is situated. Traditionally and Thomistically, our will is part of a rational soul that is the body's form and renders physically effective freely chosen action.

Accordingly, a voluntary First Cause causes both voluntary and natural second causes that are ontologically prior to, and have more potency than, mere natural causes they cause. Thomas invites the idea that, from an awareness of ourselves as *potent* voluntary causes, we may infer "Necessarily if there are such causes *(S)*, there is an *omnipotent* voluntary First Cause *(F)*": $S \Rightarrow F$. In arguing for our awareness of "both sight and its object" in the *De Anima* 425b, for instance, Aristotle provided a basis for Augustine's notion that we are undeniably aware of ourselves as well as for Thomas' position that we are *analogous* to a personal Creator.

The relation to a Creator bears on modern politics, noted shortly, and on recent science. The latter is evident in E. Farhi's "Is it Possible to Create a Universe in the Laboratory by Quantum Tunneling?".[19] In being analogous to the *creatio ex nihilo,* the tunneling brings to mind the dependence of a universe on "voluntary first-cause scientists". The question of their creating a universe, precisely not *ex nihilo,* raises a question of who created them. Since our free will, bodies, and knowledge are dependent second-causes of which we are incontrovertibly aware, our analogical relation as *Ss* to *F* pertains to $S \Rightarrow F$. Even if its transposition from $\sim F \Rightarrow \sim S$ is disputed, there may be a relevant symmetry:

If *S* bears a relation (*R*) to *F*, then *F must*
bear it to *S*

Let *R* read "is analogical to" for specifying the following

[19] E. Farhi *et al,* in *Nuclear Physics B,* B 339 (1990) 417-90.

notions: It is impossible for "F is analogical to S" (F_n) to be false when "S is analogical to F" (S_n) is true, where second creators in S_n and a First Creator in F_n may both know of the analogical relation. The relation expresses our image of a Creator in terms of such things as a rational capacity for knowledge. And in this case the knowledge includes the notion that the relation of a Creator to us may be mutually known.

The knowledge is connected ontologically to mutual awareness wherein $S_n \Rightarrow F_n$ may be recast "$S_o \Rightarrow F_o$". Thus although symmetry bears on notions or concepts, there is no *a priori* reason why some cannot have ontological import. For example, a law of nature such as the non-conservation of parity, held by physicist O. R. Frisch to permit an asymmetrical domain of Nature,[20] does not avert symmetrical notions of knowledge that bear on a human nature of lived awareness.

Apart from awareness there can be no modally intelligible talk of knowledge. And the knowledge of which we are aware is related to our *being* (S_o). The import of S_o may be captured by noting that to *be* aware of knowing we exist as second-cause creators, who are analogical to a First Creator, is to be aware of knowing there is the Creator. And thus by a material as well as formal analysis in which symmetrical concepts are not imposed *a priori* on our existence, there is a methodological rationality of reasoning from our existence to the Creator: It is impossible the statement is false that "There is a First Creator aware of knowing his analogical relation to us (F_o)" when the statement is true that "There are second creators aware of knowing their analogical relation to the First Creator (S_o)": $S_o \Rightarrow F_o$. Taken with S_o, second-cause personal creators may infer a personal First-Cause Creator in F_o:

$$(S_o \Rightarrow F_o) \wedge S_o \;/\; \therefore F_o$$

20 See O. R. Frisch, *Atomic Physics Today* (NY: Basic Books, 1961), pp. 145, 146, referenced in R. Puligandla's *Quantum Theory: Examination of the Copenhagen Interpretation* (New Delhi: Sterling Publishers Ltd., 1977), pp. 20, 21.

Why was there not a more anthropic "$S \Rightarrow F$" for a modus-ponens argument? Professor Copleston spoke of "F", in regard to *believers*, as being identified with God. Was Thomas sensitive to *nonbelievers* for whom an analogical Creator might be too close for comfort to a personal God? A mean between extremes, typical of Aristotle, may find novel voice in his epistemological mediation between nonbelievers and believers.

Thomas follows up, more than Aristotle, with Aristotle's own anthropic methodology in exactly the foregoing senses. And he expresses modalities of astonishing import. There are modal scientific necessities and an awe of a First Cause in virtue of its being omnipotent, eternal, and that on which the cosmos or world depends. The specter of a world without the Cause was admitted by several distinguished philosophers. Charles Frankel lamented an "alienation and disorientation" and Bernard Lonergan warned of meaning becoming "so debased and values so disregarded that... [the] world collapses and there follows the experience of nothingness."[21]

"Nothingness" raises the specter of there being no world. *Pari passu* there would be no world as a truth-condition for "truth". Taken with an allowance for only contingent and logically necessary truth, a contingent possibility of there being no "truth"contrasts to the truth of a conditional wherein if there is no eternal first-cause Creator, there is no world. There would be nothing rather than something. The answer to why is there something rather than nothing is that there never was nothing but always something: the Creator.

The Creator is related modally to both modal scientific necessities and a morally relevant truth of science. There is the following extraordinary realization: Thomas strengthened a realist construal of scientific truth by metaphysico-modalities. Anti-metaphysical modern philosophies undercut scientific truth, often held to be ideal knowledge, as well as any truth in morality and politics.

---

21 See Michael Novak's *Experience of Nothingness,* back cover.

Chapter 11

Afterword: The Alternative in Our Society

Thomas' modal reasoning was expressed, in a tradition arising out of Aristotle and Augustine, in the context of a metaphysical science of science. The latter contrasts to an alternative epistemic ideal of modern science that, since the Enlightenment, has fostered the disregard of a Creator with tragic political consequences. The latter were the outcome of ideas.

Enlightened ideas were expressed archetypically in d'Holbach's *Systeme de la nature* (1770). With the idea "Man is unhappy because he is ignorant of Nature," this work presaged Baron Comte's positive social-science stage of history and the pretensions to a historico-scientific view of Karl Marx. Marx exploited the prestige of physics by proclaiming: The "physicist... makes experiments. What I have to examine... is the capitalist mode of production."[1]

Marx's tacit comparison of productive *persons* to *point particles* of physics, raises timely questions about genuine science. In beginning with how *are* our nature and Nature as truth-conditions for "truth," a methodological rationality of science is at odds with pseudo-scientific ideologies that begin with political "truth".

---

[1] K. Marx, *Capital*, in *Marx*, Ed. by Allen Wood (NY: Macmillan Publishing Co., 1988), p. 205. And all other scientific ideas are mere expressions "of the dominant material relations," according to Marx's *The German Ideology* (*Ibid.*, p. 101).

The oddity is acknowledged by political scientist L. T. Sargent. He traces the initial problem to an "enlightened" optimism, prevalent among liberal to left-wing intellectuals, that all problems can be "solved by... reason and science".[2] While fascism arose in reaction even if it copied institutions of the totalitarian Left, as the *Geheime Staatspolizei* (Gestapo) was patterned on the Soviet secret police *Narodny Kommisariat Vnutrennikh Del* (NKVD), a liberal to Left optimism resulted in a view of Marx's political science as the ultimate *physical* science of science. The latter supposedly explicated even the "event" of modern physics.

Extremes of Ideological Truth

It needs to be said that a pseudo-physics of Marxist ideals may have flourished intellectually, after World War II, because the Soviets did not undergo war-crime trials as did the fascists — due to their belated allied alliance and despite a prewar occupation of Poland, mass murder, and gulags. The latter are related intellectually to the view's determinism.

Despite its dialectic, the determinism rendered incoherent any common-sense morality. Much more was there obliterated a "classless" moral nature of persons to which political institutions should be responsive. The new "science" explained deterministically all history as a manifestation of matter as well as the material origin of scientific truth itself.

Is it surprising that scientific "truth" was itself a casualty in a post-cold war politicization? When Michel Foucault said that "each society has its regime of truth, its 'general politics' of... true and false statements,"[3] which became a tenet of a science-oriented postmodernism, was not his own sentence uttered in a societal context? May it not have been incoherently both true and false in different political regimes? With recognition that affirmative answers are forcefully articulated

---

2 L.T. Sargent, *Contemporary Political Ideologies* (Ill: Dorsey Press, 1987), p. 116.

3 Michel Foucault, "Truth and Power," published shortly before he died of AIDS, *The Foucault Reader*, Ed. by A. Fontana and P. Pasquino (NY: Pantheon Press, 1984), p. 74.

elsewhere,[4] we note that the view that political power determines "truth" is scarcely either a warmed over relativistic Greek Sophism or some synthesis of Nietzsche and Marx. They attacked any objectivity except for their own extreme notions that "truth" originates in a creative will-to-power and in determined ideas of ruling classes.

As beneficial as it may have been in some ways, the Enlightenment cannot be disjoined from a contempt, of Marx and Nietzsche, for a medieval tradition. For example, Nietzsche likened this tradition to a "prison" shattered by Kant's Copernican revolution: "The pre-copernican prison... would be something retarded and regressive unless it is merely a bad joke".[5] Finding its apex in Thomas, the tradition was derided by radical epistemologies and by enlightened thought which, fueled by Kant's mediation between Rationalism and Empiricism, undercut the faith of many religious believers.

Kant's mediation led to a false dichotomy between the extremes of empirical (*synthetic*) and logically necessary (*analytic*) truth. In the course of events, many theologians were influenced by a Verification Principle which, notes Hilary Putnam, became a virtual "public norm of reasonableness".[6] Though medieval modalities would be unverifiable pseudo-statements by this naive norm of reasonableness, reason and faith would be complementary in a Thomistic Christianity that posits "truth" between the extremes. Sadly, the extremes led to politicized religious denominations.

Many denominations, since the 1960s, have become increasingly *conservative* in relinquishing a strong role of reason or *liberal* in questioning "unverifiable" articles of

---

4 See Newton-Smith's *The Rationality of Science*, p. 35, which shows such a relativism to be logically incoherent if not trivial. And R. Kearney's *Modern Movements in European Philosophy* (England: Manchester University Press, 1981), p. 8, links post-modernism, in a similar way, to ex-Marxian deconstructionists whose political agenda is to "subvert... definitive knowledge."

5 F. Nietzsche, *The Will to Power*, Tr. by W. Kaufmann (NY: Random House, 1968), p. 417.

6 H. Putnam, "Philosophers and Human Understanding," *Scientific Explanation*, Ed. by A. Heath (Oxford: Oxford University Press, 1981), pp. 99-120.

traditional faith. A political polarization has been widely followed in the news media as well as in scholarly books. The books include L. P. Baradat's *Political Ideologies* and *The Politics of Prayer* edited by Heleb Hitchcock.

After a survey of notable scholars, Baradat notes that mass ideology is unique to modern societies rooted in the Enlightenment's "rationalist tradition" and in "empirical learning... [as] the only source of knowledge."[7] And in "Words, Words Everywhere — And not a Thought to Think" in the other work, it is noted that modern knowledge led to detachments: our bodies from an understanding of human nature, our nature from Nature, and Nature itself from a "quality of real experience."[8]

Falsification in Mainstream Philosophy of Physics

In recognizing that physics has been an epistemic model for general science in the mainstream philosophy of science, it is important to expand on experiential detachments from Nature. They have a pre-enlightened basis, bearing on physics, in a Cartesian dogmatism and Humean skepticism that went unabated by Kant. Paradoxical demands that scientific truth conform *a priori* to rational norms and be derived empirically from sense experience *(i)* miss the point of empirical testing and *(ii)* undercut ascriptions of truth to universal laws or theories.

True empirical generalizations are not derived inductively and are, in any case, transposed into theoretical ones. Precisely, without recourse to Kant, Popper construed theories or laws as something akin to uninducted bold conjectures. He sought their falsification, as opposed to verifiable truth, by a material-implication reasoning.

An inability to know if theories are true when predictions are true, which reflects the reasoning's limitation, was superseded by its supposed allowance for knowing theories are false when predictions are false. But the falsification was challenged by Lakatos' notion of "research programmes" whose

---

7 L. P. Baradat, *Political Ideologies* (NJ: Prentice Hall, 1991), pp. 6-8.

8 Joyce Little, "Words, Words Everywhere — And not a Thought to Think," *The Politics of Prayer*, Ed. by H. Hitchcock (San Francisco: Ignatius, 1992), pp. 3-23.

auxiliary hypotheses may ameliorate predictive anaomalies. In any event, falsification did not afford theoretical truth. And Popper's "Inverse-Probability" (I-P) Thesis, in which improbable predictions afford a probable truth of theories, appealed surreptitiously to the induction he rejected on the basis of Hume's anti-inductivist "critical thought".

The thought was exacerbated by Kant's *Critique.* Scientific inquiry presupposes "all events are caused" and moral inquiry "all persons are free". Held to be truth-valueless, the metaphysical presuppositions erase the epistemic integrity of the inquiries. Even if they are said to afford truth for "practical purposes",[9] the truth rests on what is not strictly known to be true. And again, even before the "I-P" and "K-K" Theses — the latter specifying metaphysical knowledge for knowledge in science, there were tacit metaphysical appeals by Marx to determinism and by Nietzsche to our will, which supposes freedom, for explaining the origins of "truth".

Failure to answer "Whence comes truth if not from metaphysics?" has led to revitalized ideas of both a Nietzschean *Uebermensch* who freely creates "truth" and Marxian dominating classes that cause or condition it. The inadequacy of modern philosophy and its philosophy of science to account for "truth", exemplified by Popper's mere falsificationism and a dogmatic as opposed to methodological rationality, did not discourage radical ideological answers.[10]

Indeed, Professor Ofelia Schutte notes that disavowals of new ideological connections of Marx to Nietzsche stem from embarrassment over Nietzsche's association with fascism. She adds that, in point of fact, a *Nietzschean Marxism* flourishes with ideas of a class will-to-power to create "religion, morality, [and] philosophy... as instruments of oppression."[11]

---

9 Cf. I. Kant, *Critique of Pure Reason*, Tr. by N. K. Smith (NY: Macmillan & Co., 1965), p. 25: "knowledge... is possible *a priori*, though only from a practical point of view, to pass beyond... all possible experience."

10 See chapter 2 for Sir Karl Popper's criticism of an "anti-rational atmosphere".

11 O. Schutte, "Nietzsche... A Case of 'Nietzschean Marxism' in Peru," *Social Theory and Change* 14 (1988) 71-71.

And her insight raises the troubling question of why many contemporary philosophers have behaved obsequiously as their universities suffered a politicization.

Anti-Metaphysical Ties to Ideology

Why did many philosophers behave slavishly during the rise of political ideology in universities? Besides their own knotty epistemological problems, modern philosophy and ideology may have an anti-metaphysical "blood" that is thicker than a metaphysical "water" of ancient-medieval philosophy; in particular, a philosophic theology of Thomas. That is, if mass ideology is unique to modern society and was influenced by modern philosophy, this philosophy has more in common with ideology than with the pre-modern tradition.

This tradition is revolted against by a modern philosophy which is no less zealously anti-metaphysical than Marx and Nietzsche. Nietzsche's entanglement with the philosophy of science is clearly noted, for instance, in the seminal *Structure of Scientific Theories*: Relativistic "Weltanschauungen analyses", including the influential analyses of Feyerabend and Kuhn, are heir to a "philosophical tradition which includes Nietzsche... Accordingly, it conveniently can be viewed as a kind of neo-Kantian pragmatic position". [12]

This position rejects objective truth in favor of possibly conflicting and inconsistent historical worldviews (*Weltanchauungen*), created by Kuhnian scientific *Uebermenschen* (Supermen), of the caliber of an Aristotle or Newton or Einstein. Since physics is held to be an epistemological ideal of knowledge, such "super scientists" shape alternative ways of viewing the world that include the sciences, literature, theology, and morality. The pioneering feminist Kathryn Parsons notes, in linking Nietzsche to Kuhn's paradigms, that "A moral, social, and scientific paradigm (the paradigm of a conceptual scheme) is not merely something through which we see the world. It is something which *shapes the facts* of the human world."[13]

---

[12] Suppe, *The Structure of Scientific Theories*, pp. 126-127, fn. 258.

[13] Parsons, "Nietzsche and Moral Change," p. 186, emphasis added.

On this ideological scenario, which not only influences women's studies but an inevitable "men's studies", gender interpretations shape facts about our nature. Any objective empirical datum, which might count against the studies, is excluded a priori. "A generation after women's studies began... roots on American campuses," notes an interview with social-science professors, "scholars scrutinizing the world through the *prism of gender* have come full circle to integrate the study of men, or what it means to be male."[14]

Comparable to a *prism* that interprets the world, Kant's a priori cognitive structure was superseded by post-structural — and equally metaphysical — worldviews of sexual orientation, race, and class. In Marxian terms equally indebted to Nietzsche, "Eurocentric historical epochs" were inevitably seen as having in common dominating heterosexual white males. In the case at hand, "Each gender is left to the mercy of society to define who they become" and the studies, in holding a relativism where truth-claims may be both true and false in different societies, is to explore "diverse expressions of masculinity across race, class, and sexuality."[15]

The foregoing reflects a "Nietzschean Marxism" further illustrated by Professor Parsons' contributions to women's studies. Nietzsche and Kuhn's philosophy of science are celebrated since they are relativistic and consistent with

---

14 B. Dobbin, "New on Campus: Men's Studies Programs," *Associated Press*, 20 July 1997, emphasis added. Inspired by women's studies, the new program may further induce confusion about gender identity. On the one hand, men would not want to be traditional since "We're up for murder, rape, drug abuse, violence, aggression. Obviously, we don't know who we are since we're having all these problems." On the other, the problems are not ameliorated by less sexist roles, either, of "trying to raise a family, trying to get the job done, often with much fear and uncertainty." What "male" and "female" mean in such studies follows in the post-structural wake of a Kantian "a priori structure of mind" the thesis of which was itself metaphysical because it was synthetic a priori. That is, the thesis that "All persons have an a priori structure of mind" is itself assumed a priori, for explicating various judgments, and is synthetic since the concepts "persons" and "a priori structure of mind" have different meanings. Precisely, in this sense, the thesis is not analytic (logically true) and, in not following from experience a posteriori, is not empirically true.

15 *Ibid.*

elements of Engels' communism.[16] The epistemic mutation may be eschewed by most seasoned scholars even if they also eschew metaphysics. But the mutation is inordinately popular at lower levels of academia and is the root of political correctness.

Emergence of Relativistic Weltanschauungen

Political correctness reminds us that notions held to be "correct" may be empirically false which *should*, but is not allowed to, falsify them. And the politics has led to a questionable multiculturalism. Despite its being peddled to raise our consciousness for tolerance, a relativism has surfaced with strife between *Weltanschauungen* (worldviews) for power to determine what is "correct": liberated theological views against white-male ones, male against female views, Western versus Eastern views, modern as opposed to post-modern views and so on.

It needs to be stressed that any neutral standpoint on struggles between the worldviews is a chimera. For it is clear that, among other things, a white-male dominated theology and traditional views are to be overcome by a liberated theology and relativistic post-modernism.

For instance, J. W. Murphy and J. M. Choi argue that post-modernism exposes "racism and cultural domination... [by examining] 'foundationalism'... at the core of the Western intellectual tradition... shown to foster a metaphysics of domination."[17] A "true pluralism is possible once foundationalist approaches to knowledge... are set aside."[18] To set aside foundationalism is to reject any bedrock fact for objective knowledge. It is not knowledge or truth that becomes possible but a "true pluralism". And the latter amounts to a euphemism for multiculturalism.

---

16 K. Parsons, "Nietzsche and Moral Change," p. 185, fn. 12. She laments that, from a pragmatic standpoint, Engles' *The Origin of the Family, Private Property, and the State* was "too late for our revolutionaries to use (1884), and even if it had not been, it was far too politically radical."

17 *Cf.* J. W. Murphy and J. M. Choi, *Postmodernism, Unraveling Racism, and Democratic Institutions* (CT: Greenwood Publishing, 1997) Quotes are taken from their own GPG Book-News report (June 1997).

18 *Ibid.*

In addition to revealing a need of academic relativists to continually change words whose meanings come to be either questioned or discredited, there is a dilemma of metaphysics. Given that metaphysics and metaphysico-modalities have not been foundations of knowledge — but rather a metaphysics of domination, how can one "knowledge" supersede another that does not itself presuppose a metaphysics of either domination or nondomination? The former incurs the race-gender strife in question and the latter the question of how a nondominated metaphysics is known to be true: How can a truth-valueless *synthetic a priori* metaphysics be avoided, rooted in Kant, that is a source of both relativistic modern ideologies and postmodern philosophies in the first place?

The philosophies are imposed even on a physicist's domain of Nature. Physicist and philosopher of science A. Chalmers responds: "The natural world does not behave... in one way for males... another for females, in one way for Western cultures... another for Eastern cultures."[19] Cultures are not truth-conditions for scientific truth. Nature is the condition on pain of rendering inexplicable the universal scientific laws which afford such things as airplane flights from Eastern to Western countries. Similarly, moral norms rooted in a conscience of our nature or in Nature's God can alone tenably explain such things as international war-crimes trials for ethnic cleansings. Presumably, even academic postmodernists would be embarrassed to relegate the norms to a relativism of ethnicity or culture.

A denial that cultural worldviews determine how are human nature and Nature is an important point, even if its *patency* eludes relativists, because the denial bears on Thomas' naturalistic ethics. The ethics permits inferences from scientific descriptions of our nature to how it ought to be fulfilled, without committing a Naturalistic Fallacy, as well as allows inferences to virtues that do not both obtain and not obtain in different cultures at the same time or at different times in the same culture. The Fallacy, if anything, pertains to a modern secular ethics which has not addressed modalities.

---

[19] A. Chalmers, *Science and Its Fabrication* (Minneapolis: University of Minnesota Press, 1990), p. 112.

This brings us back to a neglect of modal inferences to a Creator who made human nature and Nature how they ought to be. Confusion of the inferences with a truthless metaphysics does not merely bear on a K-K Thesis since there is a meta-physico-modal knowledge underlying scientific knowledge.[20] The confusion bears on a modern anti-metaphysics which has replaced scholarship with ideological activity.

Under a euphemistic name of "activism", anti-metaphysics has proceeded jointly with a proselytized cultural relativism. The latter has had a secular-liberalizing impact on Judeo-Christianity. The activists nurture its politicization by *ad hominem* attacks on traditional belief as "right wing" and by incurring a backlash of believers. In understandably defending their belief, many of them may throw out a "baby" of trendy confused philosophy with a "bath water" of religio-scientific traditions that have been developed through the ages. What we have witnessed is "a tragic polarization," notes theologian Brian Harrison, in which "liberals and conservatives often forget... the words of St. Thomas".[21]

Renewed attention to Thomas' words may strengthen both Protestant and Catholic theologies, if not general philosophy itself, by raising troubling questions about the modern philosophy and philosophy of logic that have influenced the theologies. Modal reasoning is not merely compatible with them. The reasoning fortifies theology by being tied to the sciences which, since the Enlightenment, have not only come to be exploited for politicizing traditional Christian belief but misused to usurp *overlapping* beliefs of Christians as well as of Jews and Muslims.

---

20 Given that the Thomistic notion "All events have voluntary and natural causes" is known modally, the skepticism of a K-K Thesis is rendered moot: If "skepticism is to be avoided, the exploitation of... 'causal' regularities in obtaining a posteriori knowledge must not require prior knowledge of those regularities...". See Suppe, *The Structure of Scientific Theories*, p. 722. Moreover, in admitting only of natural causes expressed by Kant's causal principle, the regularities undercut coherent scientific truth itself. A causally determined reality would be the truth-condition for possibly inconsistent truth-claims.

21 Brian Harrison, O.S., "The Crisis in Eucharist Faith," *Adoremus Bulletin* (November 1995) 8.

Appendix I

A Suggested Kripkean Modal Interpretation

With a proviso that this interpretation of Thomas' conditional needs exploration and leaves many technicalities to proper sources, an irrelevance of whether or not it is in S4 might be articulated by David Lewis' notion of conditionals in a general context of Kripkean modal logic. The latter involves the specification of a relation of *which* worlds are possible relative to which others. Initially, let us recall St. Thomas' notion that "It's impossible there is a world of second causes (~*S*) when there is no First Cause (~*F*)." Here, second causes refer to experienced phenomena. And the impossibility may be recast as a necessity: Necessarily if ~*F*, then ~*S*.

Thomas' reasoning prior to this conditional might comprise a reductio wherein its denial leads to an absurdity: *S* when ~*F* involves an infinite regress of second cause phenomena that, itself being a phenomenal totality, may either cause itself or come into existence out of nothing. In taking the conditional necessarily if ~*F*, then ~*S* with *S*, we infer *F*. Having recalled these points, a version of the Lewis conditional[1] seems to be a

---

[1] The following part of this paragraph and the next paragraph are indebted to my correspondence with Professor John Nolt, May 10, 1999. I am responsible for any misstep in this account of the reasoning. See Professor Nolt's discussion of the Lewis conditional in his chapter "Kripkean Modal Logic," in *Logics* (Belmont,CA: Wadsworth Publishing Co., 1997), pp. 351-356. Also see M. Kremer's "'If' is Unambiguous," *Nous* 21 (1987) 199-217, noted by Nolt.

tenable approach for formalizing the argument. Among other things, there are truth conditions which exclude vacuous truth as well as conditions that do not require specifications of the sort of possibilities which are intended. The antecedent of the conditional does that because we are not to consider all practically, or physically, or logically possible worlds but rather all worlds most like our own in which the antecedent is true. The essential point is that the conditional "If there is no first cause, then there is no second cause" is true if and only if both of the following conditions obtain:

1. There must be some world *u* in which there is no first cause, and
2. There is no world at least as possible relative to the actual world as *u* is in which there is no first cause but there is a second cause.

If these conditions obtain and there is a second cause in the actual world, then there must be a first cause in the actual world. Thus Thomas' argument is sound.

However, this version of the Lewis semantics might seem to work in a manner that St. Thomas would reject. Presumably, as specified in clause 1, Thomas would not allow for the possibility of a world *u* in which there is no first cause. And this apparent problem results from truth conditions which exclude vacuous truth.[2] The truth conditions that Lewis himself favors, which permit vacuous truth, in effect eliminate clause 1 by allowing for the truth of Thomas' conditional even if there is no possible world where there is no first cause.

But although Lewis' semantics might appear to be best for interpreting Thomas' argument, because we need not admit of a world in which $\sim F$ is true, his "argument becomes a kind of trivial inference from the necessary existence of God as a first cause of all possible worlds."[3] Consequently, we may accept the outlined version of the Lewis conditional, pending further

---

[2] See Nolt, *Logics*, p. 351, fn. 1.

[3] Nolt, correspondence, May 10, 1999.

exploration of various alternatives, with the thought kept in mind both that this version does nevertheless allow for the inference to a first cause and that there are contexts for modal reasoning which may sometimes elude categories of modal systems such as S4.

Appendix II

Interpreting an Equivalence

A defense of Thomas' premise $\sim F \Rightarrow \sim S$, considered in chapter 9, might be put another way in terms of an equivalence $(\sim F \Rightarrow \sim S) \Leftrightarrow [S \Rightarrow (\sim F \Rightarrow \sim S)]$. Sentential-logic truth tables may be inapplicable to modalities that are not truth functional. But a *logical possibility* of the falsity of both conditionals by material implication, as opposed to strict or logical implication, is illustrated in the third row for present purposes:

$(\sim F$	$\Rightarrow$	$\sim S)$	$\Leftrightarrow$	$[S$	$\Rightarrow$	$(\sim F \Rightarrow \sim S)]$
Ft	T	Ft	T	t	T	T
Ft	T	Tf	T	f	T	T
$\mathbb{T}$f	$\mathbb{F}$	$\mathbb{F}$t	T	$\mathbb{t}$	$\mathbb{F}$	$\mathbb{F}$
Tf	T	Tf	T	f	T	T

Initially, consider the logical possibility that there may obtain "S when $\sim F$" as well as "$\sim(\sim F \Rightarrow \sim S)$ when S". The possibilities do not merely raise the question of why they should either exclude modal impossibilities or have epistemic priority over them. How can either philosophers of science or scientists allow for scientific knowledge and deny the modal impossibilities?

In beginning with Thomas' conditional "$\sim F \Rightarrow \sim S$" as the impossibility that "S when $\sim F$", we note that it underlies the behavior, if not claims, of scientists. Scientists presuppose prima facie an invariable sequence of causally *dependent* regularities for applying theories to physical systems at present times for either predictions or postdictions of them at future or past times. For example, the conditional would exclude an uncaused cosmos within which, say, universes come into and go out of existence in terms of collapses and black-hole fluctuations.

Such an uncaused cosmos would be anomalous to astrophysical theories because their application, to either its cause or past state in light of its present structure, involves the presupposition of an invariable sequence of causal regularities. That is, for the regularities to be invariable, there would have to be the notion that all phenomenal causes are caused. And this would mean that the caused causes are insufficient to cause themselves and that , in exactly this sense, they are dependent. Precisely, to say that dependent phenomena *beg modally* for an extra-phenomenal uncaused cause is to say that all phenomenal causes, of either an infinite or a finite series, are invariably subject to scientific inquiry.

And to say the inquiry presupposes an impossibility there are the dependent phenomena $(\sim S)$ when there is no extra-phenomenal first cause $(\sim F)$, *e.g.* $\sim F \Rightarrow \sim S$, is to say $(\Leftrightarrow)$: "Necessarily if there are the phenomena $(S \Rightarrow)$, then this fact is impossible $(\sim S)$ when there is no extra-phenomenal first cause $(\sim F)$". That is, $S \Rightarrow (\sim F \Rightarrow \sim S)$. In short, this is to say $(\sim F \Rightarrow \sim S) \Leftrightarrow [S \Rightarrow (\sim F \Rightarrow \sim S)]$.

Moreover, phenomena have causal relations that find expression in scientific modalities. The difference between what is modally begged for and what is *logically possible* becomes clear by the fact, for instance, that for physicists to acknowledge the *logical possibility* of constructing a periodically functioning machine doing more work than energy supplied from outside is not to acknowledge the *modal possibility*. Various kinds of modal impossibilities are given priority. The impossibilities are so patently true that they are frequently unexpressed until challenged by empirical claims. These claims range from periodically functioning machines doing more work than energy supplied from outside to ordinary human conception without normal fertilization.

Further, the impossibilities are supposed surreptitiously by scientists who view their inquiries as yielding approximately true theories: It is impossible that the theories are not approximately true in given domains when they afford systematically successful predictions and manipulations of phenomena in the domains.

Moreover, there is a broad spectrum of domains from physics to astronomy to geology to biology and to pioneering applications of bio-medical research, as already discussed, that are characterized by modal scientific truths. They also set parameters for what are reasonable empirical claims. For example, the claim is excluded that Peter came into the world without prior biological processes.

In denoting S as "S_m" for second modal relations that set parameters for empirical claims, consideration of the equivalence $(\sim F \Rightarrow \sim S_m) \Leftrightarrow [S_m \Rightarrow (\sim F \Rightarrow \sim S_m)]$ may start with an analysis of $\sim F => \sim S_m$: Necessarily if there is no first cause related causally to second modal relations that set parameters for empirical claims $(\sim F)$, there are not second modal relations that set the parameters $(\sim S_m)$: $\sim F \Rightarrow \sim S_m$.

To affirm S_m, by a modal analogy of modus tollens, is to affirm F. And to affirm $\sim F$ when allowing for S_m, by appealing to logical possibility, is to allow paradoxically for the possibility that $\sim S_m$. *Pari passu* $\sim S_m$ does not merely permit the logical possibility of persons coming into the world without prior biological processes. It allows in a general manner for things both coming into existence from nothing and going out of existence in the absurd sense of simply disappearing.

Such an allowance would not merely conflict with a perennial and fundamental experience of human nature and Nature, which is Thomas' point. The allowance, at least in terms of a "lived logic", would be viewed by virtually all modern laypersons as well as by philosophers and scientists as more seriously in error than mere empirical falsities. And this point holds *even if* the seriousness of the error is ignored in a modern epistemology that commits, prima facie, a false-dichotomy fallacy by recognizing only empirical and logically necessary truth (between which there lie true modalities).

In recalling W. H. Walsh's criticism of modern anti-metaphysical philosophies in his *Metaphysics* — not to mention Wittgenstein's insights in *On Certainty*, there are modal errors that lie in seriousness between logically impossible and empirical ones.[1] A modal error of supposing that a quarter in one's pocket materialized out of nothing would be *less* serious than supposing that it both is and is not in one's pocket. At the same time, the modal error would be *more* serious than in mistakenly supposing that the quarter was in one's right pocket when in fact it was in the left.

Having noted such forgotten common-sense modalities, we note that scientific ones would involve prior modal relations. For instance, metamorphic phenomena related modally to marble, which set parameters for reasonable empirical claims — say by excluding claims that specify or imply that marble was produced when there were not the phenomena, are related modally to prior phenomena and they to other prior ones and so on. In other words, metamorphic phenomena of great heat and pressure cannot obtain without phenomena causing them and ones causing those and so forth.

In sum, to say there is a series of modal relations related modally to others (S_m), setting parameters for empirical claims, is to say both that S_m cannot obtain when it is not related modally to F and that the modal relation between F and S_m also sets parameters for empirical claims. These claims would include either that S_m obtains when F does not or that it is false that $\sim F \Rightarrow \sim S_m$ when S_m. That is, where S_m denotes a second-cause modal series insufficient to cause itself, the conditional is equivalent to: Necessarily if there is a second-cause series

---

1 W. H. Walsh, *Metaphysics: An Exposition and Defense of a Controversial Branch of Philosophy* (NY: Harcourt, Brace & World, 1962), pp 154-58. Walsh's categorial-modality is complemented by Wittgenstein's *On Certainty* (NY: Harper & Row, 1972): "There are cases where doubt is unreasonable, but others where it seems logically impossible... [with] no clear boundary between them [42e]." *E.g.* "someone came into the world without parents... [59e]." Again, the latter can be interpreted in a manner consistent with developing bio-physical technologies such as cloning.

insufficient to cause itself *($S_m \Rightarrow$)*, this fact is impossible ($\sim S_m$) when there is no first cause sufficient to cause it ($\sim F$). That is, $S_m \Rightarrow (\sim F \Rightarrow \sim S_m)$.

And in recalling that S_m involves modal scientific relations which set parameters for reasonable empirical claims and that $\sim F \Rightarrow \sim S_m$ virtually guarantees a causal dependence supposed by scientific inquiry,[2] affirming that $\sim F \Rightarrow \sim S_m$ cannot be false when S_m is true should be no bitter pill for either philosophers of science or scientists to swallow.

Here, the point is not that $\sim F \Rightarrow \sim S_m$ has no evident modal truth inferred from experience. Given a metaphysical view of science as ideal knowledge— itself a *synthetic a priori* view without known truth from a modern standpoint, the point is that a valid inference of $\sim F \Rightarrow \sim S_m$ from the modal scientific relations S_m may foster a more sober-minded consideration of the conditional's truth.

---

2 An impossibility there are modal relations when they are unrelated to *F* does not conflict with Peter Geach's distinction between *(i)* God brought it about that (E*x*) (*x* is an A) and *(ii)* (E*x*) (God brought it about that *x* is an A). That is, *(ii)* means that God makes into A an entity presupposed to his action and *(i)*, in not meaning this, is understood to express Thomas' point that "in creation the thing created does not undergo change..." Geach adds that "in creation there is no real relation of God to the creature." This may simply mean that the creature, if created by *F*, is not related causally to *F* in the sense that persons are related analogically to *their* creations. The latter are created by persons when they change pre-existing entities into new things as opposed to *F qua* God who may also create things without their being related to previous ones. See Geach's *God and the Soul* (NY: Schocken Books, 1969), p. 83.

Appendix III

Faith and Reason of a Religious Philosopher

The evolution of politicized truth in communist and fascist societies into a political correctness in open societies, with an apologetics in modern philosophy, has become apparent to even ordinary citizens. In this respect, "The Philosopher Pope" is a striking editorial by virtue of noting unparalleled societal pathologies which stem from philosophies rooted in the Enlightenment.[1] It illustrates both that heirs of the Enlightenment have largely ignored a mission of pursuing truth and that one of the most eminent religious philosophers gives expression to a harmony of faith and reason, apart from any religious denomination, which alone promises a return to cultural sanity:

> The powerful encyclical issued last week by Pope John Paul II will not be the subject of earnest debate like his calls for social justice in a materialistic world. Nor is it likely to ignite religious arguments over his unbending positions on church doctrine or his stands against abortion, euthanasia and women in the priesthood. Instead, this pronouncement... is a missive from a religious philosopher to the world's intelligentsia. It is a plea for an end to the separation of faith and reason and an argument against the "philosophy of nothing,"...

---

[1] J. Lelyveld, Executive Editor, "The Philosopher's Pope," *The New York Times*, p. A30..

Written for the bishops of the Catholic Church, the decree called "Fides et Ratio" (Faith and Reason), argues that these two strains of human endeavor need not exclude each other in the pursuit of truth. Science and rational thought do not wipe out the exploration of "the fundamental questions which pervade human life..."

Within the church, the encyclical undoubtedly signals a return to the study of philosophy in the priesthood, a subject that has often been down-played in recent years in favor of sociology and other more current fields. The Pope also wants to bring the questions of morality back into the pursuit of science — an issue that at once seems as medieval as Galileo and as modern as Hiroshima [given the influence of Hume's Naturalistic Fallacy on modern philosophy wherein from no scientific description of what *is* the case can a prescription be validly deduced about what *ought* to be the case]...

Finally, the decree argues against the modern philosophies — post-Enlightenment realism,[2] Marxism,

---

2 This post-Enlightenment realism was noted in the last chapter as one beholden to interpretations of modern Rationalism and Empiricism wherein there arose a false dichotomy of empirical and logically necessary truth. The necessary truth is trivial in not bearing on reality. And empirical truth could not only *not* be ascribed to universal scientific laws and theories, which portend to describe what reality is really like, but is excluded from a causal principle presupposed by scientific inquiry. Since the inquiries of science and ethics presuppose the principles that all events are caused and persons are free, which were each relegated to a metaphysics not known to be true, philosophy gave rise to political ideologies. The ideologies responded to the question "whence comes truth if not from metaphysics" with a tacit metaphysics that appealed to either dominating classes (Marxism) or Uebermenschen (fascism) who caused or created "truth"; hence the ensuing troubles of this "century, including World Wars, the Holocaust, the Stalin years" and so forth. Intellectually, these tragedies are rooted in the disregard of an alternative modal truth by which a more liberal causal principle is true and permits the inference to a first-cause Creator. In creating the world, this Creator affords reasoning from scientific descriptions of either a Nature or human nature — that is as it ought to be — to how it ought to be fulfilled morally and institutionalized politically.

nihilism — that see the search as the goal, so that "everything is fleeting and provisional [relativistic]," ...John Paul... knows these philosophers and has argued with their tenets. As a Pole, he also understands the troubles of his century, including World Wars, the Holocaust, the Stalin years, the gaudy capitalism that replaced Communism.

This encyclical may not convert the nihilists, but it promises that even if reason does not always lead to belief, a reasoning person like a scientist or a philosopher can still be a believer.

Bibliography

Abelson, Raziel, ed. *Ethics and Metaethics*. New York, NY: St. Martin's Press, 1963.

Adams, E. M. "The Accountability of Religious Discourse." *International Journal of Religion* 18 (1985) 3-17.

Addis, Laird. The "Memorial Minutes" on Gustav Bergmenn. *Proceedings of the American Philosophical Association* 61 (1987) 165.

Albert, E. M. et al. *Great Traditions in Ethics*. New York: D. Van Norstrand Co., 1988.

Aquinas, St. Thomas. *Summa Theologica*. In *A Summa of the Summa*. Ed. by Peter Kreeft. San Francisco: Ignatius Press, 1990.

Aquinas, St. Thomas. *De Veritate in Communi. In Quaestiones Disputatae*. Ed. by R. M. Spiazzi, O.P. 2 Volumes. Rome-Turin: Marietti, 1954.

Aquinas, St. Thomas. *De Principiis Naturae*. Ed. by J. J. Pauson. Fribourg: Societe Philosophique, 1950.

Aquinas, St. Thomas. *De Regimine Principum*. Tr. by G. B. Phelan. London: Sheed and Ward, Institute of Mediaeval Studies, 1938.

Aquinas, St. Thomas. *Summa Contra Gentiles*. Tr. by the English Dominican Fathers. London: Burns Oates, 1924.

Aronson, Ronald. *Jean-Paul Sartre: Philosophy in the World*. London: Verso, 1980.

Augustine, Saint. *On Christian Doctrine*. Tr. by D. W. Robertson, Jr. New York: Macmillan Publishing Co., 1988.

Augustine, Saint. *On Free Choice of the Will*. Tr. by A. S. Benjamin and L. H. Hackstaff. New York: Macmillan Publishing Co., 1985.

Augustine, Saint. *Confessions*. Tr. by R. S. Pine-Coffin. Middlesex, England: Penguin Books Ltd., 1984.

Augustine, Saint. *City of God*. Tr. by G. Walsh, S.J., D. Zema, S.J., G. Monahan, O.S.U., and D. Honan. New York: Doubleday & Co., 1958.

Ayer, A. J. "What I Saw When I was Dead." Reprinted by Arrangement with *The London Sunday Telegraph* in the *National Review* 40 (1988) 38-40.

Ayer, A. J. *Language, Truth, and Logic*. New York: Dover Publications, 1952.

Bair, A. "Parents, Schools, and Values Clarification." *The Wall Street* Journal, 12 April 1982.

Baradat, Leon. *Political Ideologies: Their Origin and Impact*. New Jersey: Prentice Hall, 1991.

Barnes, Hazel. *An Existentialist Ethics*. Chicago: The University of Chicago Press, 1978.

Basler, R. P., ed. *The Collected Works of Abraham Lincoln*, Vol. V: 1861-1862. New York: Rutgers University Press, 1953.

Bell, A. A. and Allis, J. B., eds. *Resources in Ancient Philosophy*. London: The Scarecrow Press, Inc., 1991.

Berge, P., Pomeau, Y., and Vidal, C. *Order Within Chaos*. New York: John Wiley & Sons, 1984.

Beverly, Elizabeth. "Liberals Must Confront the Conservative Argument." *The Chronical of Higher Education*, 1 Nov. 1989, A 52.

Boyd, R. "Realism, Approximate Truth, and Philosophical Methods." *The Philosophy of Science*. Ed. by D. Papineau. New York: Oxford University Press, 1996.

Briggs, David. "Faith Gives Hope in Troubled Times." *The Associated Press*, 16 April 1995.

Chroust, Anton-Hermann. *Aristotle: New Light on His Life and on Some of His Lost Works*, Volume II. Notre Dame, IN: University of Notre Dame Press, 1973.

Clark, Joseph T., S.J. *Conventional Logic and Modern Logic. With a Preface by W. V. Quine, Harvard University*. Woodstock, MD: Woodstock College Press. Issued by the Office of the Secretary of The American Catholic Philosophical Association, Washington 17, D.C., 1952.

Clarke, N., S.J. Professor Emeritus of Philosophy at Fordham University, professional correspondence, 20 February 1997.

Codevilla, A. *The Character of Nations: How Politics Makes and Breaks Prosperity, Family, and Civility*. New York: Harper Collins Publishers, 1997.

Connell, R., Professor Emeritus of Philosophy at the University of St. Thomas, professional correspondence, 20 August 1996.

Copleston, Frederick, S.J. *A History of Philosophy 2: Medieval Philosophy II*. New York: Image Books, 1962.

Crabb, L. J. *Effective Biblical Counseling*. Grand Rapids, Mich: Zondervan Publishing, 1977.

De George, Richard. "Theological Ethics and Business Ethics." *Journal of Business Ethics* 5 (1986) 424.

Deutsch, David and Lockwood, Michael. "The Quantum Physics of Time Travel." *American Scientist* 270 (1994) 68-74.

Devaney, R. L. *An Introduction to Chaotic Dynamical Systems*. New York: Addison-Wesley Publishing Co., 1988.

Edel, A. "Romanel Lecture." *Proceedings of the American Philosophical Association* 60 (1987) 823-840.

Farhi, E., Guth A., and Guven, J. "Is it Possible to Create a Universe in the Laboratory by Quantum Tunneling?" *Nuclear Physics B*, B339 (1990) 417-490.

Frisch, O. R. *Atomic Physics Today*. New York: Basic Books, 1961.

Gamut, L. T. F. *Logic, Language, and Meaning*: Volume II: Intensional Logic and Logical Grammer. Chicago: The University of Chicago Press, 1991.

Garrett, L. "Drugs Barely Keep Ahead of Bacteria." *Newsday-KY Enquirer*, 13 March 1995.

Gingerich, Owen. "Space, Time, and Beyond: The Place of God in the Cosmos." *Gross Memorial Lecture*. Valparaiso, IN: Valparaiso University Press, 1993.

Ginsberg, M. "Nietzschean Psychiatry." *Nietzsche: A Collection of Essays*. Ed. by R. C. Solomon. New York: Anchor Books, 1973.

Goldberg, Steven. "Bob Dylan and the Poetry of Salvation." *National Forum* LXIX (1979) 20.

Gonsalves, M. A. *Fagothy's Right and Reason: Ethics in Theory and Practice*. London: Merrill Publishing Co., 1989.

Greenberger, M., and Zeilinger, A. *Fundamental Problems in Quantum Theory*: A Conference Held in Honor of Professor John A. Wheeler. NY: Annals of the New York Academy of Sciences, Volume 755, 1995.

Haack, Susan. *Deviant Logic, Fuzzy Logic: Beyond the Formalism*. Chicago: The University of Chicago Press, 1996.

Haack, Susan. "Concern for Truth: What it Means, Why it Matters." *The Annals of the New York Academy of Sciences* 775 (Summer 1996) 57-63.

Haack, Susan. *Philosophy of Logics*. London: Cambridge University Press, 1978.

Harries, Karsten. "Heidegger as a Political Thinker." *The Review of Metaphysics* 24 (June 1976).

Harrison, Rev. Brian, O.S. "The Crisis in Eucharist Faith." *Adoremus Bulletin* (November 1995) 8.

Hauke, Manfred. *Women in the Priesthood? A Systematic Analysis in the Light of the Order of Creation and Redemption*. Tr. by D. Kipp. San Francisco: Ignatius Press, 1988.

Heelan, P. "Quantun Mechanics and Objectivity." *Problem of Scientific Realism*. Ed. by E. A. MacKinnon. New York: Appleton-Century, 1972.

Heidegger, Martin. "Deutsche Studente." *Freiburger Studentenzeitung*, 3 November 1933. Reprinted in *Guido Schneeberger. Nachlese zu Heidegger*, Dokumente zu seinem Leben und Denken. Bern, 1962.

Heidegger, Martin. *An Introduction to Metaphysics*. Tr. by R. Manheim. New York: Doubleday & Co., 1961.

Hitchcock, Helen Hull. Ed. *The Politics of Prayer: Feminist Language and the Worship of God*. San Francisco: Ignatius Press, 1992.

Hook, S. "Modern Knowledge and the Concept of God." *Critiques of God*. Ed. by P. Angeles. New York: Promethius, 1977.

Hurley, P. J. *A Concise Introduction to Logic*. Belmont, CA: Wadsworth Publishing Co., 1991.

Jensen, R. V. "Sigma Xi National Lectures 1994-1995." *American Scientist -- Sigma Xi Scientific Research Society* 81 (1993).

John Paul II, Pope. *Crossing the Threshold of Hope*. Tr. by V. Messori. NY: Alfred A. Knopf, 1994.

Jones, W. T. *A History of Western Philosophy: The Medieval Mind,* Vol. II. New York: Harcourt Brace Jovanovich, 1969.

Kahane, K. and Tidman, P. *Logic & Philosophy*. New York: Wadsworth Publishing Co., 1995.

Kaufmann, Walter. *Existentialism From Dostoevsky to Sartre.* New York: NAL Penguin Inc., 1975.

Kearney, Richard. *Modern Movements in European Philosophy.* Manchester, England: Manchester University Press, 1994.

Kenny, Anthony, ed. *The Oxford History of Philosophy.* Oxford: Oxford University Press, 1994.

Korner, Stephen. *Experience and Theory.* London: Routledge & Kegan Paul, 1969.

Kreeft, Peter. *A Shorter Summa: The Essential Philosophical Passages of St. Thomas Aquinas' Summa Theologica Edited and Explained.* San Francisco: Ignatius Press, 1993.

Kreeft, Peter. *Fundamentals of The Christian Faith: Essays in Christian Apologetics.* San Francisco: Ignatius Press, 1988.

Leftow, Brian. "A Modal Cosmological Argument." *International Journal For Philosophy of Religion* 24 (1988) 159-188.

Levison, A. "A Century Awash in Blood -- Spotlight: Genocide." *The Associated Press,* 17 September 1995.

Lindberg, David C. *The Beginnings of Western Science: The European Scientific Tradition in Philosophical, Religious, and Institutional Context, 600 B.C. to A.D. 1450.* Chicago, Illinois: The University of Chicago Press, 1992.

MacPhee, D. "Directed Evolution Reconsidered." *American Scientist -- Sigma Xi Scientific Research Society* 81 (1993) 554.

Malcolm, Norman. *Ludwig Wittgenstein: A Memoir. With a Biographical Sketch by G. H. von Wright.* New York: Oxford University Press, 1984.

Marcus, Ruth Barcun. *Modalities: Philosophical Essays.* New York: Oxford University Press, 1993.

Moody, E. A. *Studies in Medieval Philosophy, Science, and Logic.* Berkeley: University of California Press, Center for Medieval and Renaissance Studies, 1975.

Monk, Ray. *Ludwig Wittgenstein*: The Duty of Genius. London: Penguin Books Ltd., 1990.

Murphy, J. W. and J. M. Choi. *Postmodernism, Unraveling Racism, and Democratic Instututions*. Westport, CT: Greenwood Publishing Group, 1997.

Newton-Smith, W. H. *The Rationality of Science*. London: Routledge and Kegan Paul, International Library of Philosophy, 1980.

Novak, Michael. *The Experience of Nothingness*. New York: Harper & Row, 1971.

Nolt, John. *Logics*. Belmont, CA: Wadsworth Publishing Co, 1997.

O'dea, Thomas et al. "Religion in the Year 2000." *Philosophy Looks to the Future*. Prospect Hts, Il: Waveland Press, 1985.

Pitkin, Hanna. *Wittgenstein and Justice*. Berkeley: University of California Press, 1974.

Plumwood, Val. "Nature, Self, and Gender." *Environmental Philosophy*. Ed. by M. Zimmerman, J. Callicott, G. Sessions, K. Warren, and J. Clark. Englewood Cliffs, NJ: Prentice Hall, 1993.

Popper, Karl, Sir. *Quantum Theory and the Schism in Physics*. New Jersey: Rowman & Littlefield, 1982.

Puligandla, Ramakrishna. *Quantum Theory: An Examination of the Copenhagen Interpretation*. New Delhi: Sterling Publishers Ltd., 1977.

Putnam, Hilary. "Philosophers and Human Understanding." *Scientific Explanation: Herbert Spencer Lecture*. Ed. by A. F. Heath. Oxford: Oxford University Press, 1981.

Ratzsch, D. Professor of Philosophy at Calvin College, professional correspondence, 17 December 1996.

Rescher, Nicholas. "Where Wise Men Fear to Tread." *American Philosophical Quarterly* 59 (1990) 259.

Ritter, Malcolm. "The Egg Didn't Wait." *The Associated Press*, 29 September 1995.

Rohrlich, F. and Hardin, L. "Established Theories." *Philosophy of Science* 50 (1983) 603-617.

Ross, David, Sir. *Aristotle*. New York: Barnes & Noble, Inc., 1964

Rothman, M. A. *A Physicist's Guide to Skepticism*. New York: Prometheus Books, 1988.

Russell B. *The Autobiography of Bertrand Russell*. V. I. Boston: Atlantic-Little, Brown, 1967.

Sahakian, W. *History of Philosophy: From the Earliest Times to the Present*. New York: Barnes and Noble, 1968.

Saperstein, A. M. "War and Chaos." *American Scientist-Sigma Xi Scientific Research Society* 83 (1995) 548-557.

Sartre, Jean-Paul. *Being and Nothingness*. Tr. by Hazel Barnes. New York: Philosophical Library, Inc., 1956.

Schaaffs, W. Theology, *Physics, and Miracles*. Tr. by R. Renfield. Washington, D. C.: Canon Press, 1974.

Schutte, Ofelia. "Nietzsche, Mariategui, and Socialism: A Case of Nietzschean Marxism in Peru?" *Social Theory and Practice* 14 (1988) 71-85.

Sluga, H. and D. G. Stern, eds. *The Cambridge Companion to Wittgenstein*. New York: Cambridge University Press, 1996.

Smith, P. *Killing the Spirit*. New York: Penguin Books, 1990.

Stenger, V. J. "The Face of Chaos." *Free Inquiry* 13 (1993) 13-14.

Stumpf, S. E. "St. Augustine's Christian Philosophy." *Socrates to Sartre: A History of Philosophy*. New York: McGraw-Hill Book Co., 1988.

Sullivan, R. J. *Morality and the Good Life*. Memphis, TN: Memphis State University Press, 1977.

Suppe, Frederick, ed. *The Structure of Scientific Theories*. Chicago: University of Illinois Press, 1977.

Taylor, Charles. *Sources of the Self*. New York: Cambridge University Press, 1989.

Toyton, Evelyn. "The Wittgenstein Controversy." *The Atlantic Monthly* 279 (June 1997) 28-41.

Trundle, Robert C., Jr. *From Physics to Politics*. NJ: Transaction Publishers -- Rutgers, The State University, 1999.

Trundle, Robert C., Jr. "Modalidades aristotelicas de san Agustin: En memoria de Fr. Jose Oroz Reta." *Review of Augustine* (Avgstinvs) 42 (1997) 13-40.

Trundle, Robert C., Jr. "Cold-War Ideology: An Apologetics for Global Conflict." *Res Publica: Leuven Politologisch Instituut* 37 (1996) 61-84.

Trundle, Robert C., Jr. "St. Thomas and Modal Logic: On Wittgenstein's and Heidegger's Possibility." *Aquinas: International Review of Philosophy* 38 (1995) 223-248.

Trundle, Robert C., Jr. "Thomas' 2nd Way: A Defense by Modal Scientific Reasoning." *Logique et Analyse* 146 (1994) 145-168.
Trundle, Robert C., Jr. *Ancient Greek Philosophy: Its Development & Relevance to Our Time*. Avebury Series in Philosophy. London: Ashgate Publishing Co., 1994.
Trundle, Robert C., Jr. "St. Augustine's Epistemology: An Ignored Aristotelian Theme and Its Intriguing Implications." *Laval Theologique et Philosophique* 50 (fevrier 1994) 187-205.
Trundle, Robert C., Jr. "Quantum Fluctuation, Self-Organizing Biological Systems, & Human Freedom," *Idealistic Studies* 50 (1994) 269-281.
Trundle, Robert C., Jr. "St. Augustine's On Free Choice of the Will: A Prologue to 20th C. Meaning." *Avgstinvs*. Ed. by P. Merino. Madrid, Spain: Orden de Augstinos Recoletos, 1994.
Trundle, Robert C., Jr. "Applied Logic: An Aristotelian Organon for Critical Thinking in a Theoretical Context." *Philosophy in Science* 5 (1993) 117-140.
Trundle, Robert C., Jr. "Physics and Phenomenology." *New Horizons in the Philosophy of Science*. Avebury Series in the Philosophy of Science. Ed. by David Lamb. London: Ashgate Publishing Co., 1992.
Trundle, Robert C., Jr. "The Cases For and Against Theological Approaches to Business Ethics." *Laval Theologique et Philosophique* 47 (1992) 241-259.
Trundle, Robert C., Jr. "Business, Ethics, & Business Ethics: 2nd Thoughts on the Business-Ethics Revolution." *Thought: A Review of Culture & Idea* 66 (1991) 297-309.
Trundle, Robert C., Jr. Review of John W. Yolton's *Locke and French Materialism*. *The Modern Schoolman: A Quarterly Philosophical Journal* 67 (1991) 75-78.
Trundle, Robert C., Jr. "De Interpretatione IX: Problem of Future or Infinite Past Truth?" *The Modern Schoolman: A Quarterly Philosophical Journal* 59 (1981) 49-55.
Van Bendegem, J-P. Chairman, Department of Philosophy at Vrije Universiteit Brussel, professional correspondence, 5 May 1996 through 14 August 1996.
Waismann, Friedrich. *The Principles of Linguistic Philosophy*. Ed. by R. Harre. New York: St. Martin's Press, 1968.

Walsh, W. H. *Metaphysics: An Exposition and Defense of a Controversial Branch of Philosophy*. New York: Harcourt Brace & World, 1962.

Wittgenstein, Ludwig. "On Heidegger on Being and Dread." *Heidegger & Modern Philosophy*. Ed. by M. Murray. New Haven, CT: Yale University Press, 1978.

Wittgenstein, Ludwig. *On Certainty*. Ed. by G. E. M. Anscombe and G. H. von Wright. Tr. by D. Paul and G. E. M. Anscombe. London: Harper & Row, 1969.

Wittgenstein, Ludwig. *Philosophical Investigations*. Tr. G. E. M. Anscombe. New York: The Macmillan Co., 1958.

Wittgenstein, Ludwig. *Tractatus Logico-Philosophicus*. Tr. by D. F. Pears and B. F. McGuinness. Introduction by Bertrand Russell. London: Routledge & Kegan Paul, 1961.

Yavorsky, B. M. and Seleznev, Yu. *Physics: A Refresher Course*. Moscow: MIR Publishers, 1979.

Yolton, John W. *Locke and French Materialism*. Oxford: Clarendon Press, 1991.

Youssef, Saul. "Is Quantum Mechanics an Exotic Probability Theory?." *Fundamental Problems in Quantum Theory: A Conference Held in Honor of Professor John A. Wheeler*. Ed. by D.M. Greenberger and A. Zeilinger. NY: Annals of New York Academy of Sciences, V. 755, 1995.

Zinn, G. A. and Kibler, W. W., eds. *Medieval France*. Hamden, CT: Garland Publishing, 1995.

Zyskind, H. and Sternfeld, T. R. "Plato's Meno 89c: 'Virtue is Knowledge' A Hypothesis?." *Phronesis* 21 (1976) 130-134.

INDEX

A

B

Y